AF711178

Published by
Princeton Architectural Press
37 East Seventh Street
New York, New York 10003
For a free catalog of books, call 1.800.722.6657.
Visit our website at www.papress.com.

Original edition © 2009 22 Publishing, Milan
The original edition of this book was published by
22 Publishing S.r.l.
Via Morozzo della Rocca 9
20123 Milan, Italy
www.22publishing.it

This edition © 2009 Princeton Architectural Press
All rights reserved
Printed and bound in China
12 11 10 9 4 3 2 1 First edition

No part of this book may be used or reproduced in any manner without written permission from the publisher, except in the context of reviews.

Every reasonable attempt has been made to identify owners of copyright. Errors or omissions will be corrected in subsequent editions.

22 Publishing Editor: Alessandro Rocca
22 Publishing Designer: Daniela Rossi

Translator: Steve Piccolo

Princeton Architectural Press Editor: Becca Casbon

Special thanks to: Nettie Aljian, Bree Anne Apperley, Sara Bader, Nicola Bednarek, Janet Behning, Carina Cha, Penny (Yuen Pik) Chu, Carolyn Deuschle, Russell Fernandez, Pete Fitzpatrick, Wendy Fuller, Jan Haux, Clare Jacobson, Aileen Kwun, Nancy Eklund Later, Linda Lee, Laurie Manfra, John Myers, Katharine Myers, Lauren Nelson Packard, Dan Simon, Andrew Stepanian, Jennifer Thompson, Paul Wagner, Joseph Weston, and Deb Wood of Princeton Architectural Press —Kevin C. Lippert, publisher

Library of Congress Cataloging-in-Publication Data

Corbellini, Giovanni, 1959-
[Bioreboot. English]
Bioreboot : the architecture of R&Sie(n) / Giovanni Corbellini.
p. cm.
Originally published: Milan : 22 publishs.r.l, 2009.
Includes bibliographical references.
ISBN 978-1-56898-869-6 (alk. paper)
1. R & Sie (Architectural firm) 2. Architecture and biology. I. Title. II. Title: Architecture of R&Sie(n).
NA1053.R25C6713 2009
720.92'2—dc22
2009011659

Giovanni Corbellini

Bioreboot
The Architecture of R&Sie(n)

Princeton Architectural Press, New York

colonies, biotopes, and other robotic proliferations

the natural architecture

Sustainable colonies, personal and social biotopes, robotic proliferations: the natural architecture of R&Sie(n) represents a contribution of exceptional importance for the construction of a new relationship between the knowledge, techniques, and imagery of contemporary architecture and the world of nature. In our time our impact on the environment—its disturbances and changes—has become a global problem that raises fundamental questions about the way in which man inhabits this planet.

In R&Sie(n)'s architecture the pathologies of a living, suffering world become the raw materials of design, seen as a palimpsest on which to apply the commands of extreme procedures. Ironic and paradoxical premises and modes are disorienting yet firmly rooted in the concrete dimension of a feasible action that seems desirable, even necessary, once its reasons have been understood. With respect to the polished, translucent, and preferably immaterial features of much of the architecture of today, these projects appear as obstinate, abnormal mechanisms that refuse to dematerialize, unconcerned about their cumbersome, disturbing physical presence. Humors, fluids, membranes, tissues, and biomorphic structures construct disquieting images of apparatus in movement, animated by the programmed vitality of the robot, the golem, the Frankenstein's monster at-large. Every project by R&Sie(n) is like a heavy machine—a mechanism that performs a given number of functions with a certain degree of intelligence (minimal, perhaps, but decisive) based on a predetermined protocol.

The beauty of these projects, which have their own charm and brutality, is that each of them is a process, a dynamic device applied to a fragment of the Earth's crust, with the tenacity of an artificial parasite. Each building resembles an automatic animal that performs an ecologically useful function with humility and practical spirit, using every means offered by architecture and the context to achieve its results, in an apparently amoral approach of absolute utilitarianism—that of the survival instinct of living beings.

The force of these bizarre architectural creatures lies, therefore, in their capacity to make reference—with the devotion of a mechanical servant—to the processes and rhythms of natural architecture, i.e. architecture whose primary aim is the ecological and social improvement of the place in which it exists, penetrating the labyrinth of the desires, fears, and hopes of contemporary man.

Alessandro Rocca

of R&sie(n)

contents

insects, robots, electronic arms, and other devices for the architecture of

ecosophical apparatus & sc

recognition of the eroticism resulting from nature conflicts and a built-in mosquito trap for a sadomasochistic negotiation with paranoia

< mosquito bottleneck >

R&Sie(n)

with
Stephan Heinrich,
robot designer

izoid machines

recognition of the eroticism resulting from human conflicts and biomass collecting machines for a haunted house

< he shot me down >

In power games, [apparatus could be considered] relationship strategies supporting types of knowledge and supported by themselves.
—Michel Foucault, *Dits et Écrits 1954–1998*

Machines are always pretending to do more than what they were programmed to do. It's their nature. Their behavior induces in us phantasms, frustrations, and fears inspired by their ability to break free and threaten us (as the Golem did to its creator, Rabbi Loeb).

The blurriness between what they are supposed to do—as perfect, alienated, and domesticated creatures—and

fading topography and an animatronic penguin from antarctic stories

< terra incognita >

recognition of the eroticism resulting from nature conflicts and a built-in mosquito trap for a sadomasochistic negotiation with paranoia

<mosquito bottleneck>

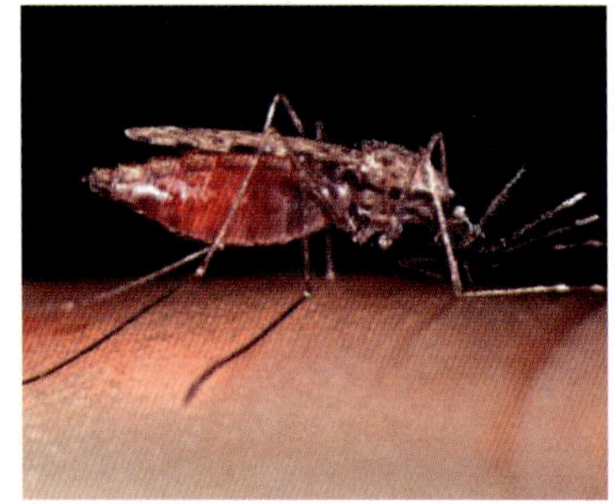

recognition of the eroticism resulting from human conflicts and biomass collecting machines for a haunted house

< he shot me down >

the anthropomorphic psychology we intentionally project on them creates a spectrum of potentiality—both interpretative and productive—that is able to re-scenarize the operating processes of the architectural field. Machines are a vector of narration, generators of rumor, and at the same time directly operational, with an accurate efficiency of production.

These multiple disorders, this kind of schizophrenia, could be considered a tool for reopening processes and subjectivities, for re-protocolizing indeterminacy and uncertainties. In this way, they become agents of blur logic, of reactive and reprogrammable logic.

As in *Alice's Adventures in Wonderland*, where Lewis Carroll used mathematics to confuse a little girl's perception, such apparatus, including "bachelor machines" (in the sense of Duchamp and Picabia) stretch a line of subjectivization to organize repetitions and anomalies (as in Gilles Deleuze's *Différence et répétition*, 1968), using and developing paradoxes that are able to recomplexify and de-alienate the edges of the truth system, in order to reinvert the logic of meaning and turn it into a vanishing point.

It seems to make strategic sense to evaluate architecture's degree of reality based on its ability to tell stories, and in this way enlarge the dimension of its physicality. In a sense, we should consider the structure itself as a fragment of a scenario—a MacGuffin—the point from which speeches, strategies, scientific protocols, and power games articulate stories and agendas.

Misunderstandings, in this sense, produce artifacts—in the "garden that forks nowhere"—and apparatus can

stochastic machine producing a time repeating, deleuzian "body without organs"—an experiment with virtual potentials

< olzweg >

fading topography and an animatronic
penguin from antarctic stories

< terra incognita >

be considered generators of ambiguities and knowledge where nonshaping protocols, protocols of emergence, contingently reveal the conditions of emission and are revealed by them, in a Situationist strategy.

R&Sie(n)

animal paganism and a power station
for a contemporary ceremony place

< hybrid muscle >

swarm intelligence, random agents,
and stochastic scripts for an
unpredictable, interactive urbanism

< swarm town >

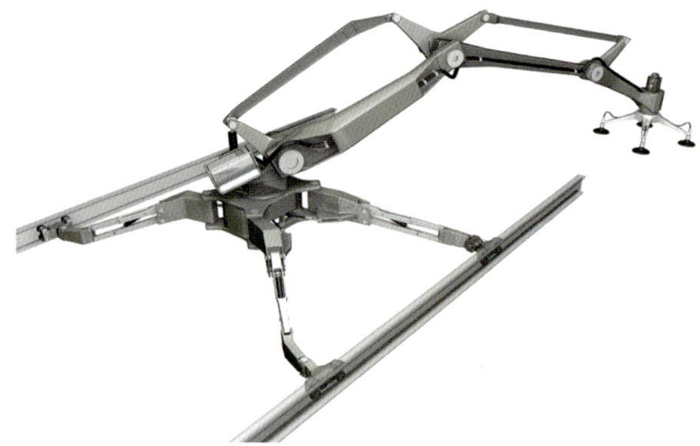

recognition of the eroticism resulting from human conflicts and biomass collecting machines for a haunted house

< he shot me down >

stochastic machine producing a repeating, deleuzian "body without organs"—an experiment with virtual potentials

< olzweg >

nostalgic-industrial transformation of wood in an icy climate for a simulacrum of tradition

< waterflux >

pollution and identity are extracted from the dirty air through an electrostatic filtering device

< b_mu/dusty relief >

anomalies: disturbed and disturbing architectures

an

Giovanni Corbellini

omalies

multiple identity

R&Sie(n) is the only architecture studio whose members avoid having pictures of themselves appear in magazines, books, and other media. The various publications usually contain a strange, androgynous figure with amber skin, which upon close examination appears to be a morphing of multiple faces, in a double—front and profile—portrait. To find out more about the people that are more concealed than revealed by this sort of simultaneous mug shot, we can look at their biographical notes, which indicate that their equally cryptic name, R&Sie(n), comes from the contraction of the names of the three present partners: François Roche, Stéphanie Lavaux, and Jean Navarro. This name is an almost-automatic recording of the composition of the studio and its respective roles and relationships, so much so that over time the name of the studio has often changed, evolving to reflect the variations of the team.[1] Somewhat like the Citroën DS (which, when pronounced in French, sounds like the word *déesse*, or "goddess"), but in the opposite direction, Roche and partners seem to be interested in the heresy of presenting themselves in a problematic way to the outside world.[2] The mutant "trademark" and media iconophobia (a recent interview in *Volume* is illustrated only with François's hand gestures) show a sophisticated manipulation of the most established dogmas of the postmodern religion of marketing, where absolutely everyone—from television pitchmen to politicians, Osama bin Laden to starchitects—pursues efficacy of penetration, in markets and minds, through the continuity and recognition factor of a brand.[3]

In an era in which individual charisma seems to have definitively taken the place of ideologies and grand narratives, this programmatic obscuring of personal imagery may seem both disturbed and disturbing. We are reminded of J. D. Salinger and Thomas Pynchon and their maniacal defense of their privacy, or the almost equally reclusive Terrence Malick (who has not granted an interview since 1973). The possible intersections with their works and those of R&Sie(n) are not lacking: the dialectic between purity and corruption in *The Catcher in the Rye*, the scientific, conspiracy-mad paranoia in the stories and novels of Pynchon, the threatening beauty of nature in *The Thin Red Line*. Nevertheless, R&Sie(n)'s oblique relationship with the public sphere certainly develops on different planes. They are surely closer, in terms of generational and political-cultural references—especially references to the Situationist International and its revolutionary actions—to the anonymity practiced by the collective of Italian writers known as Luther Blissett / Wu Ming, above all in the effects of paradoxical media overexposure. The "open source" strategy that permitted the multiplication of actions under the handle Luther Blissett (not just on the part of the authors of *Q*, but also through a vast international network of artists and social activists) is reversed in a propagation of identity whose schizoid differentiation leads to a similar spread in visibility, further stimulated by the proliferation of links offered by the internet.[4]

This is clearly an intentional disruption of the myth of the auteur, not just in opposition to the excess of personalization that continues to permeate the star system of international architecture, but also, and above all, a strategic option capable of guaranteeing the freedom of action that the self-referential fetters of a brand (or of an architectural language) and its necessities of commercial reproduction would impede. An uncertain, multiple, mutant identity becomes a fundamental prerequisite (and an eloquent signal) of an open design attitude, which cannot be pigeonholed as a particular, albeit efficacious, modus operandi. R&Sie(n) use a guerrilla tactic capable of infiltrating apparatus in which the spasmodic pursuit of novelty clashes with an equally powerful and paralyzing demand for security.

utopia

Rhetorical reversal is one of the specific devices of the avant-garde, and it is also a mechanism we can recognize in the systematic shifts in the design action of R&Sie(n).

The fact that similar procedures of inversion emerge on the delicate plane of personal identity is significant and original, especially in a discipline that bases its characterizing specificity with respect to other arts precisely on the rationality of its approaches and the efficacy of its results. One of the typical motifs in the work of Roche and partners, in fact, is the continuous, strategic use of personality disorders (and of imagery possessed by corruption and danger), which act as motors for the conceptual shifts required for the configuration of innovative, surprising scenarios. This is a way of taking the Paranoiac-Critical Method to radical extremes that makes them different from other protagonists of the disciplinary debate, whose projects rely on equally paradoxical but undoubtedly less disturbing operations.[5]

The world of architecture, though only in relatively recent phases, has also started to use the random processes previously applied in literary research (Alfred Jarry, Stéphane Mallarmé, Raymond Queneau), art (starting with Marcel Duchamp), and, above all, the "hard" sciences. In the sphere of Darwinian evolution, quantum mechanics, game theory, and multiple research projects and applications, the apparent nonsense of a probabilistic, indeterminate approach has turned out to be a very effective strategy, capable of interpreting complex situations and leading to positive intervention.

To design by "playing dice" allows architects to stay in contact with elusive, volatile realities, where the compositional control of morphological order has become, to an increasing extent, an objective that is an illusion, and has definitively lost its appeal at the same time. As a result, more and more architectural research tends to shift its fundamental nucleus from the definition of forms and the languages that organize them to the procedures capable of subtracting significant parts of the project from the control of its maker. In other words, "dice" are utilized, case by case, to give meaning to different design strategies. Among the many designers working on experiments with all kinds of random processes, Roche and partners make systematic use of the unconscious and its pathologies, connecting them with the long series of more or less mad geniuses the worlds of art and science (from Vincent van Gogh to Nikola Tesla, Alan Turing to Syd Barrett), as well as the world of architecture (such as Antoni Gaudí, prey to his mystic transport, or Adolf Loos, a suspected pedophile, or Walter Gropius, dysgraphic and a notorious sex fiend), have never stopped producing. François Roche would be perfectly capable of playing the role of the "eccentric" artist, but the obsessions that pervade his projects, in the end, become surprisingly lucid operative tools. The penetrating analysis behind them seems to be closely connected to a very French current of thought, driven by personalities and movements of different origins: the Marquis de Sade, Duchamp, the Surrealists, Jean Dubuffet, Georges Bataille, Jacques Lacan, Foucault, Deleuze and Félix Guattari, and the Situationists. This current is crossed, in the diversity of its many facets, by a strong antipositivist thrust, where individual impulses and their most radical liberation represent the fundamental components of any negotiation with reality.

In this sense, one of R&Sie(n)'s most significant projects is undoubtedly **I've Heard About**. It is literally a utopia, a research project independent of any positioning, however hypothetical, in a specific territory (an oddity for an architecture studio whose characteristic approach involves connecting with local conditions). The urban idea it proposes is a sort of upgrade of New Babylon, the ideal city where Constant experimented, fifty years ago, with the Situationist prospect of the end of labor (delegated to sketchily described underground machines) and the recreational practice of the *dérive* ("drift") as a spontaneous, necessary activity.

< multiple

utopia

The continuous reformulation of the spaces of social interaction connected with this liberating hypothesis is taken to its extreme consequences here, through a permanent self-organization that goes beyond the dialectic between predominantly fixed infrastructure and mobility/transformability of smaller partitions (a hypothesis once advanced, in a fully capitalist perspective, in the Plan Obus by Le Corbusier). The machines, no longer hidden below ground, expand, demolish, reconstruct, and modify the environment, controlled by sophisticated software capable of combining constructive logic and the demands of the inhabitants.

Mixing real possibilities with science fiction scenarios, the research of R&Sie$^{(n)}$ proceeds on different levels: it develops the protocols needed to guarantee structural solidity, studies growth algorithms, experiments with a construction system based on fiber cement, and designs the machines capable of making the city on-site. At the same time, they consider the modes of interaction between individual desires and built structure, imagining a biopolitical system of electronic interconnection (where each person can formulate his own requests in real time) functioning in integration with a prepsychic level measured by microsensors scattered in the ether, capable of chemically recording the level of metabolites and therefore gauging the amount of collective stress.

Combining "automatic" technical expertise (no longer controlled by flesh-and-blood professionals) with direct forms of "democratic" self-determination, the city of Roche and partners (if it can still be called a city) delves into the problematic issues of previous radical experiments—from Yona Friedman's to Archigram's, Superstudio's to the Metabolists'—that had already explored the megastructural theme. Nevertheless, **I've Heard About** does more than just resolve the issues that were often merely hinted at by earlier experiments. The infrastructural and disciplinal functions served by the large bridge sectors of New Babylon (or by the reticular structure of Friedman's Villa Spatiale) are transferred, here, into the immaterial component of a rigorously open-source software script that ensures the sustainability of permanent negotiation among the various actors involved. It is as if the grid of Manhattan—whose rigidity is the premise and condition for extreme variation in its three-dimensional extrusion—had become an accessible protocol that could be modified in real time, open to the interferences and retroactive effects of self-organization.

The introduction of a level of unconscious negotiation also challenges the rational apparatus shared, though with very different accents, by the radical utopias. This prepsychic aspect, a source of such extreme spontaneity that it is even ahead of our conscious will, represents the fundamental operator of the project for R&Sie$^{(n)}$, to the point of becoming the center of the **Hypnochamber**, the simulation installed in 2005 at the Musée d'Art Moderne de la Ville de Paris. In this installation, a "hypnotic chamber" immersed the visitor in an altered state of consciousness, a sort of parallel reality in which to experience the "chemical" connection between the urban organism and its continuous, "live" self-determination. **Hypnochamber** is an expression of complex individual and collective subjectivities in search of a forcefully widened terrain for cultivation, even with respect to the liberating hypotheses of Constant. There is no longer a question of establishing zones of freedom within definite limits, but of continuously revising and modifying the limits themselves, on the basis of latent, instinctive impulses: beyond the relationships of power, history, morals, and good common sense.

architects: R&Sie(n) / Benoît Durandin   year: 2005

an ideal, social, speculative, self-organized, "contour crafting" neighborhood protocol for a flat, fat, growing urban experiment

client: Musée d'Art Moderne de la Ville de Paris, France   dimensions: growing

open-source software for a

< i've heard about >

habitable organism

The VIAB device, for extrusion of fiber cement: Developed with the Robotics Research Lab of the University of Southern California, this construction machine takes its name from the terms *viability* and *variability*.

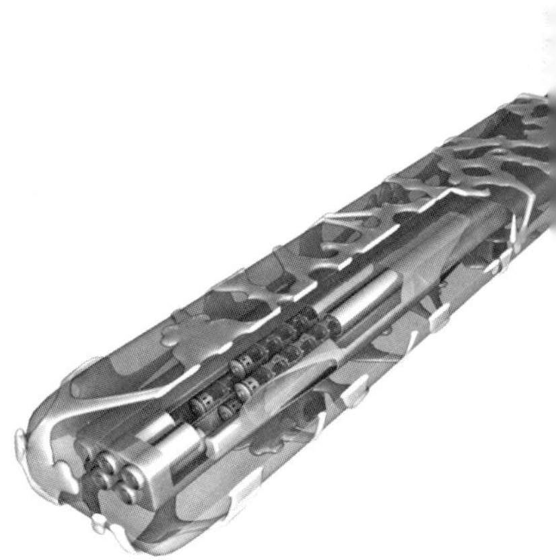

scenario

1)

Speculation about a habitable organism: This develops by means of adaptive, transitory scenarios, in which the key operational mode is uncertainty.

2)

Writing of growth scripts, open algorithms: These should be sensitive not only to human expression, but also to data such as the chemical emissions of the inhabitants.

A generative algorithm controls the static viability of the structure and interacts with the subjectivity of the residents.

$$V_{n+1}$$

$$\nearrow\!\swarrow \vec{u}_{n_i}$$

$$\begin{array}{c}\longrightarrow\\ \longleftarrow\end{array} \quad V_n = V_{n+2} = V_{n+8}$$

$$\nwarrow\!\searrow \qquad\qquad V_{n+4}$$

$$\nearrow\!\swarrow$$

$$V_{n+3} = V_{n+5}$$

$$\nwarrow\!\searrow$$

$$\vec{u}_{n_j} \qquad V_{n+6}$$

$$\vec{u}_t(q) = V\big(\rho_i(q), \varepsilon_i(q) \mid c(q)\big)$$

3)

Creation of software as an open source that can be modified and infiltrated by its entropic development

4)

Contraction and operation of a contour-crafting building machine (the VIAB) that is capable of pouring fiber-reinforced concrete: It should also reflect the constantly evolving interaction between the inhabitants and their environment.

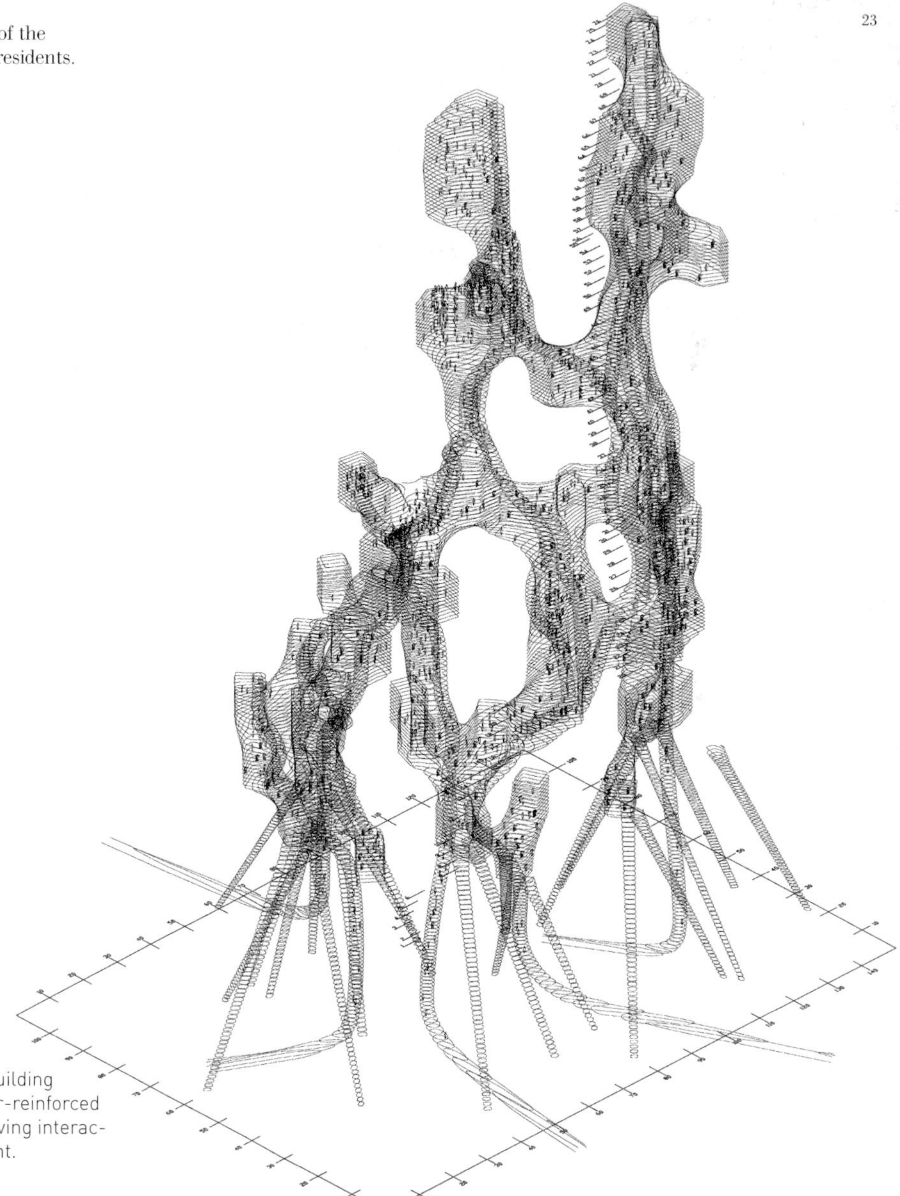

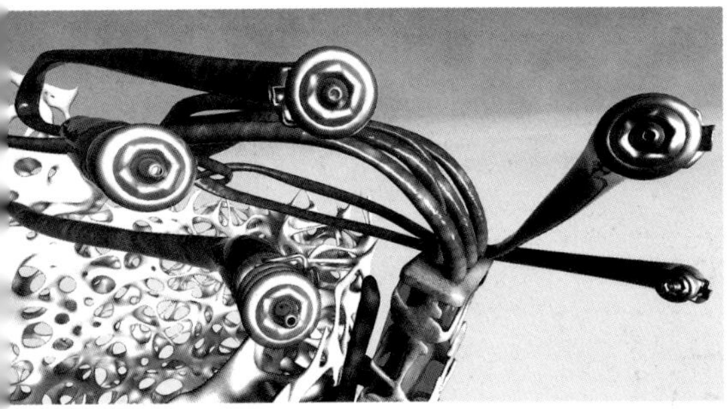

The VIAB is a sort of parasite, like the polyps that live in coral and allow it to proliferate. It was inspired by phasmids, insects that can camouflage themselves as a leaf or a stick.

I've Heard About recognizes and constructs a fragile, mutable sociality. Its growth is based on real-time negotiation with residents.

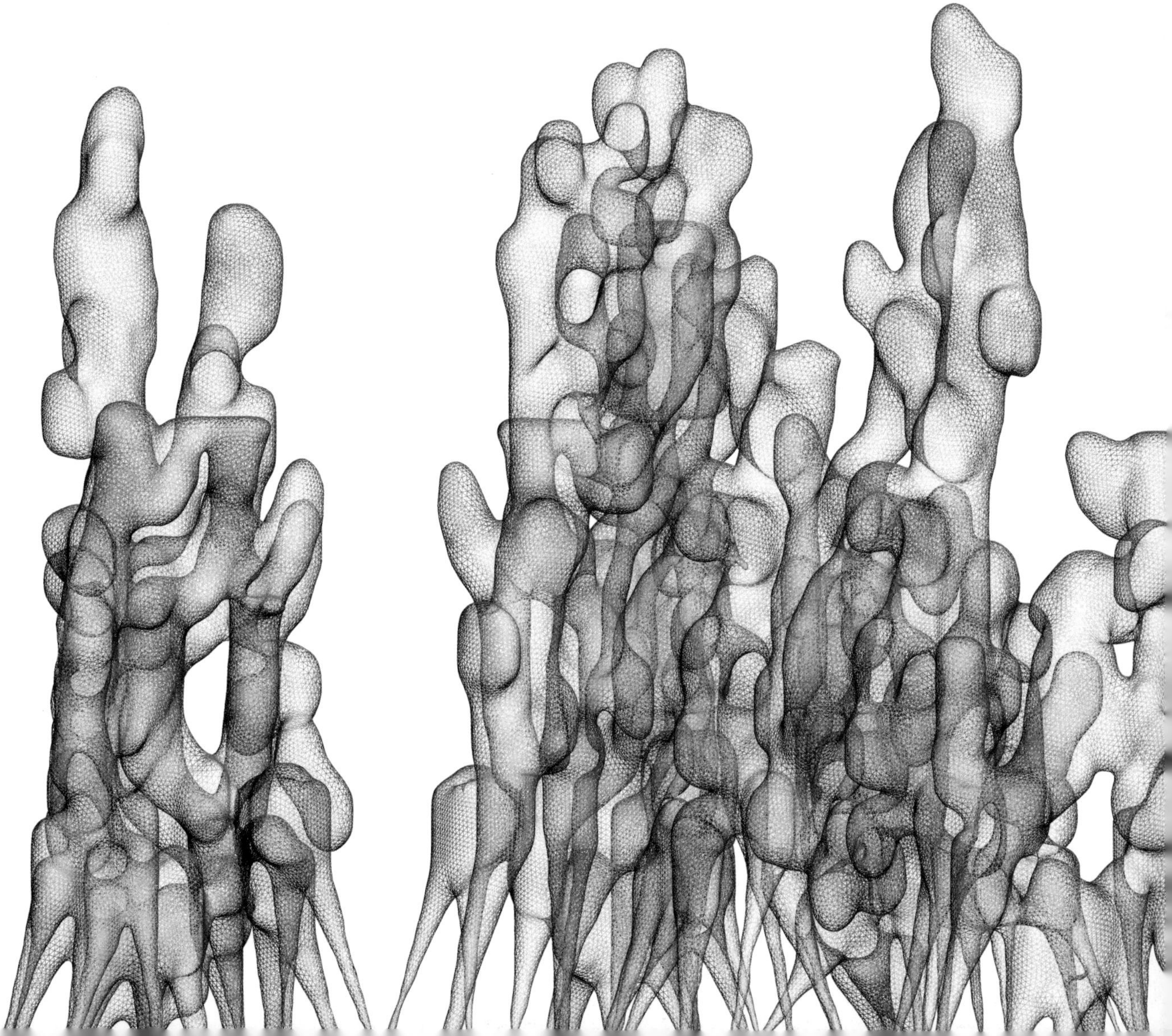

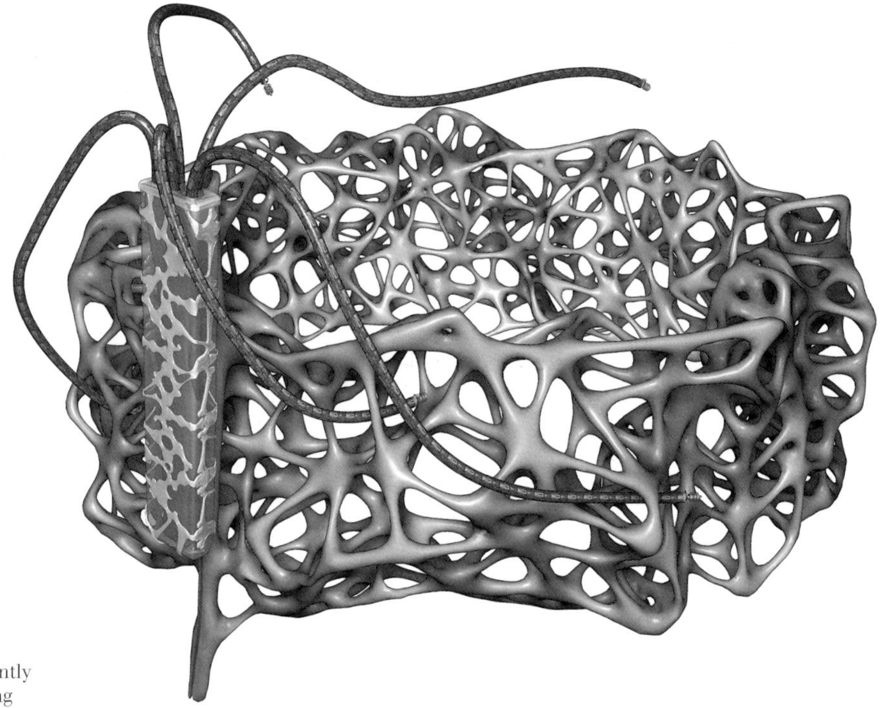

The VIAB device in action:
It is an architectural organism constantly under construction, capable of utilizing incompleteness and self-determination as its constituent parameters.

Internal colonization of the structure:
Kits made with special inflatable, self-polymerizing plastics make interstices inhabitable, providing the necessary closures, slabs, connections, units of chemical exchange, and hydroponic gardens.

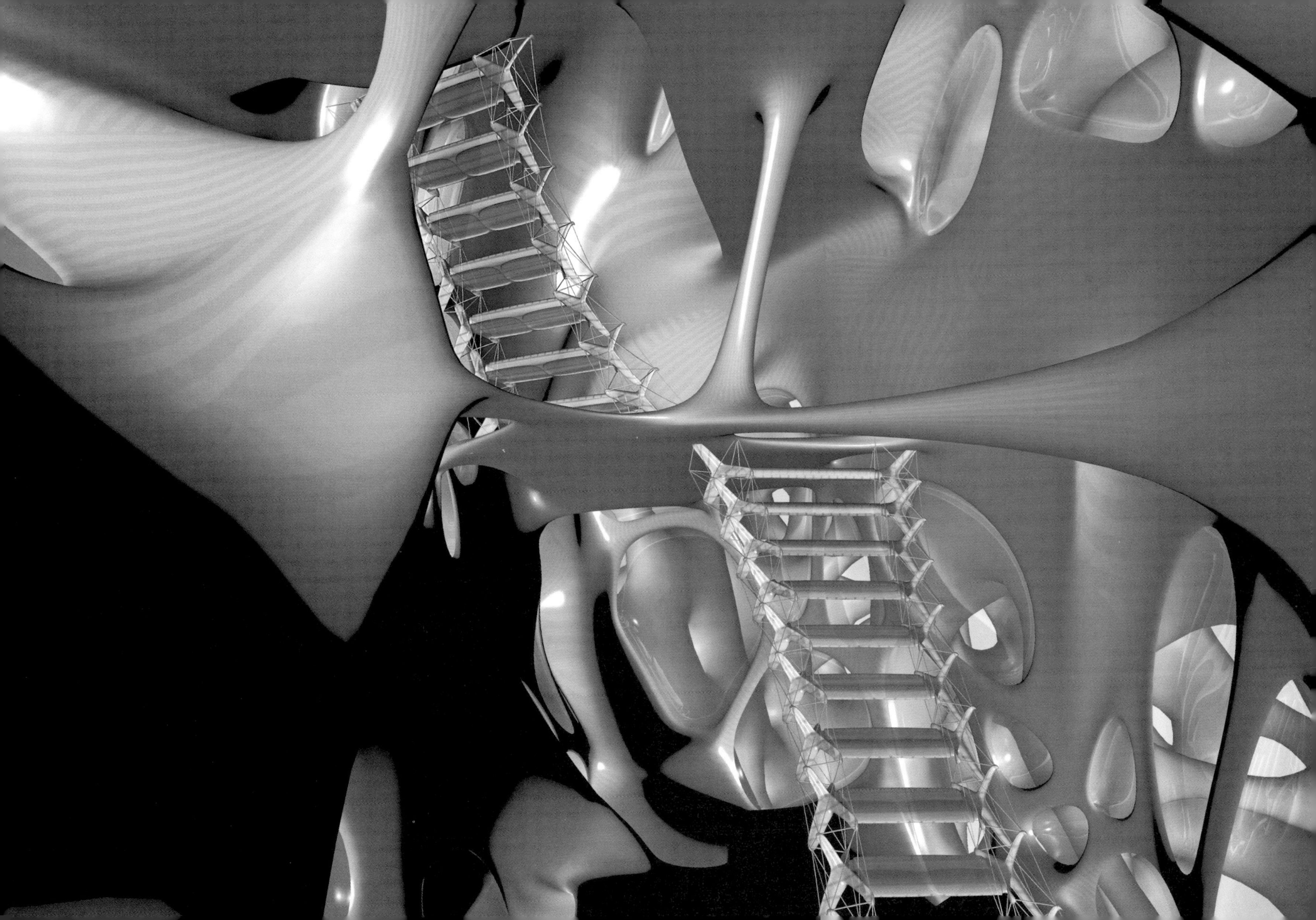

architects: R&Sie(n) / Benoît Durandin   year: 2005   location: Musée d'Art Moderne de la Ville de Paris (MAM), France

a hypnotic, individual-session, psycho-operative space for the research and exhibition of i've heard about

client: MAM   cost: $157,500 (€120,000)   dimensions: 161 ft.^{2} (15 m^2)

unconscious urban

< hypnochamber >

planning test

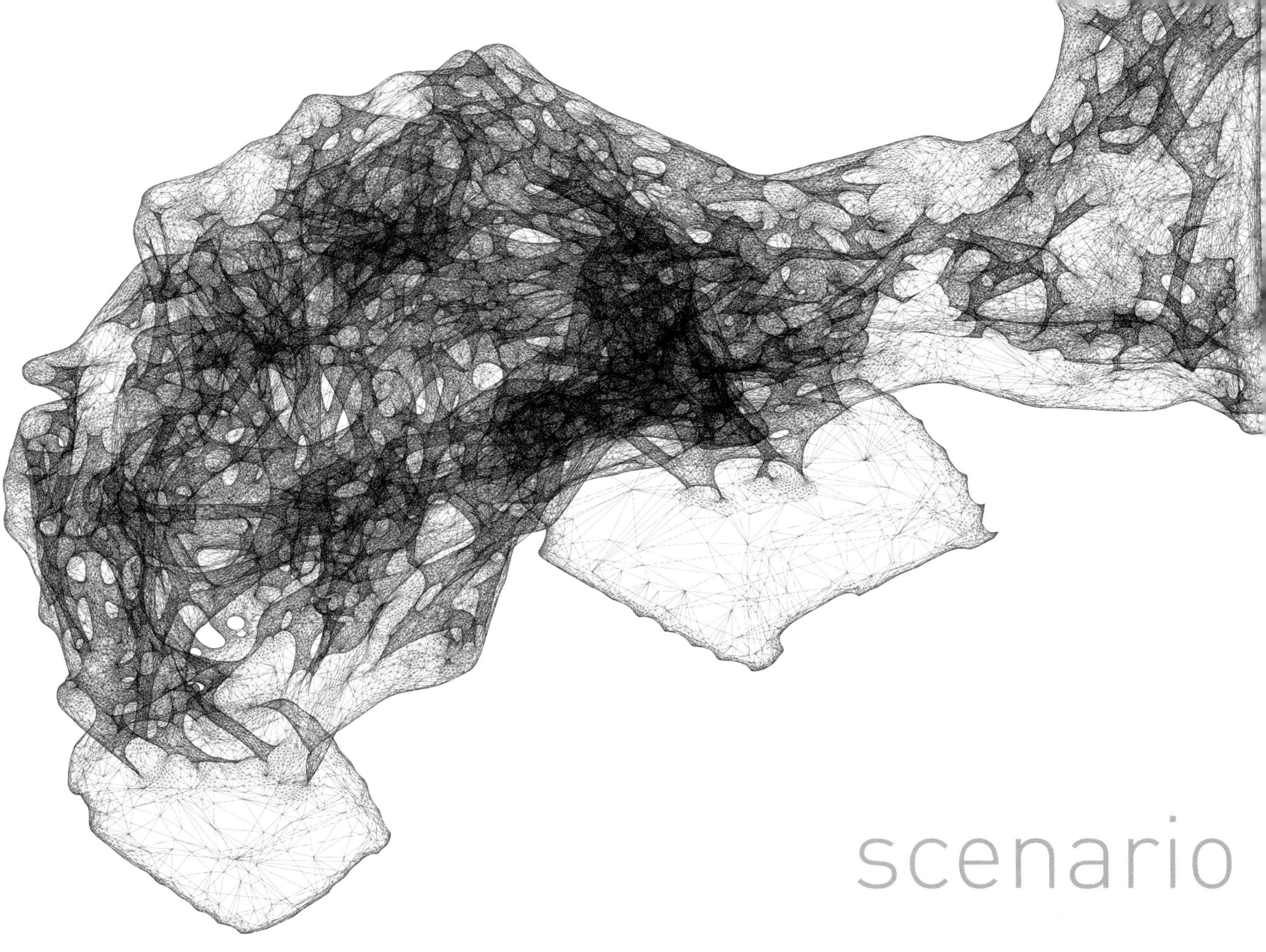

scenario

Virtual model:
Hypnochamber is an interior space, a zone of immersion in which to take part in a prerecorded hypnosis session that helps the visitor escape from the alienation of social control and experience the new condition of citizenship imagined in I've Heard About.

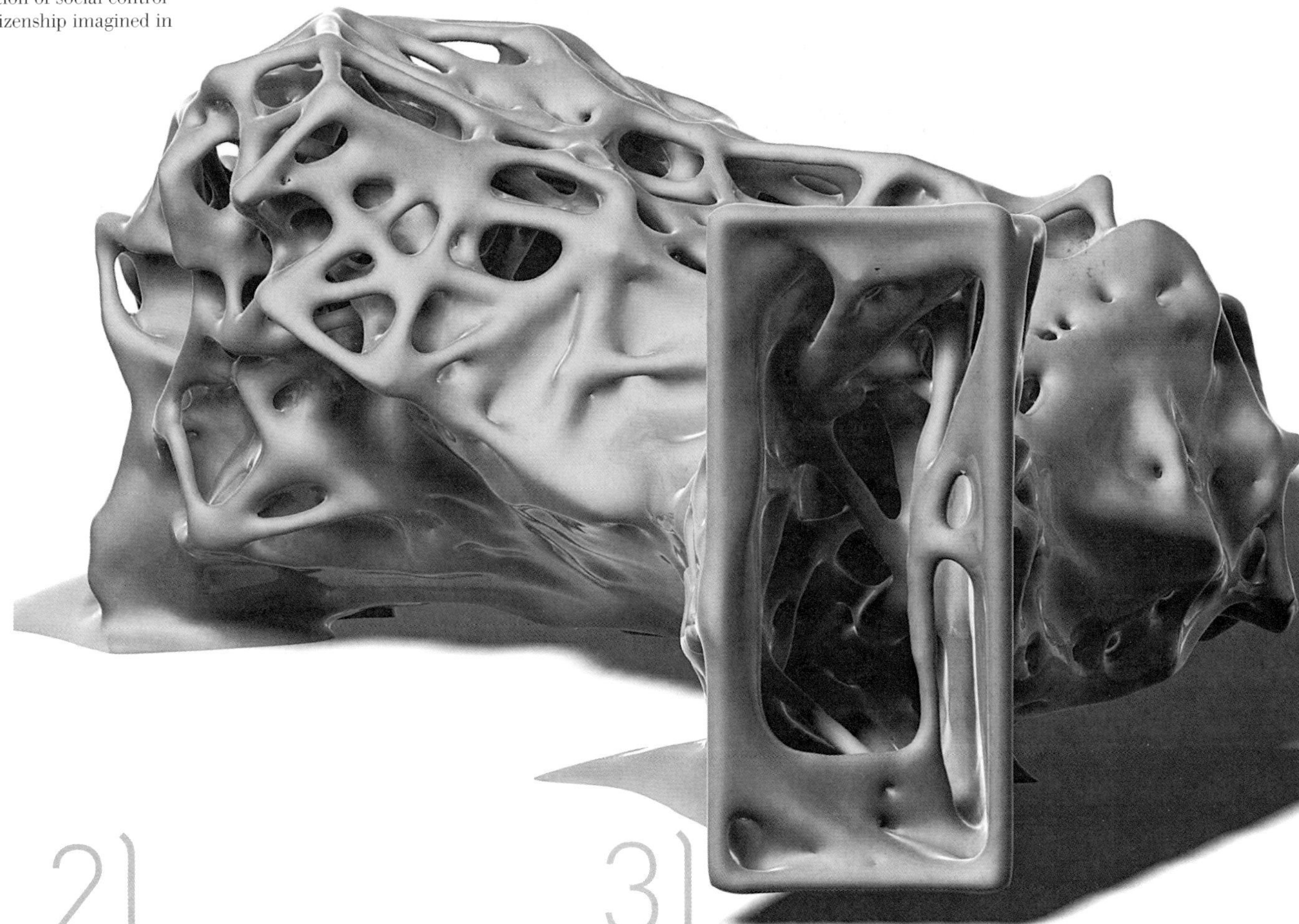

1)

Enter a "heterotopian," cognitive room.

2)

Secondly, dive into a waking dream while listening to a voice describing the urbanism of I've Heard About.

3)

Perceive yourself as a nerve ending inside the organic, self-determined growth structure.

Interior of the Hypnochamber:
How to produce disalienation and infiltrate subjectivity? The Hypnochamber is a way of constructing a personal relationship with a narrative environment. It is a door that leads to a different level where the notion of democracy is transformed into a process of self-determination.

Real prototype of a sector of I've Heard About installed in the Musée d'Art Moderne in Paris

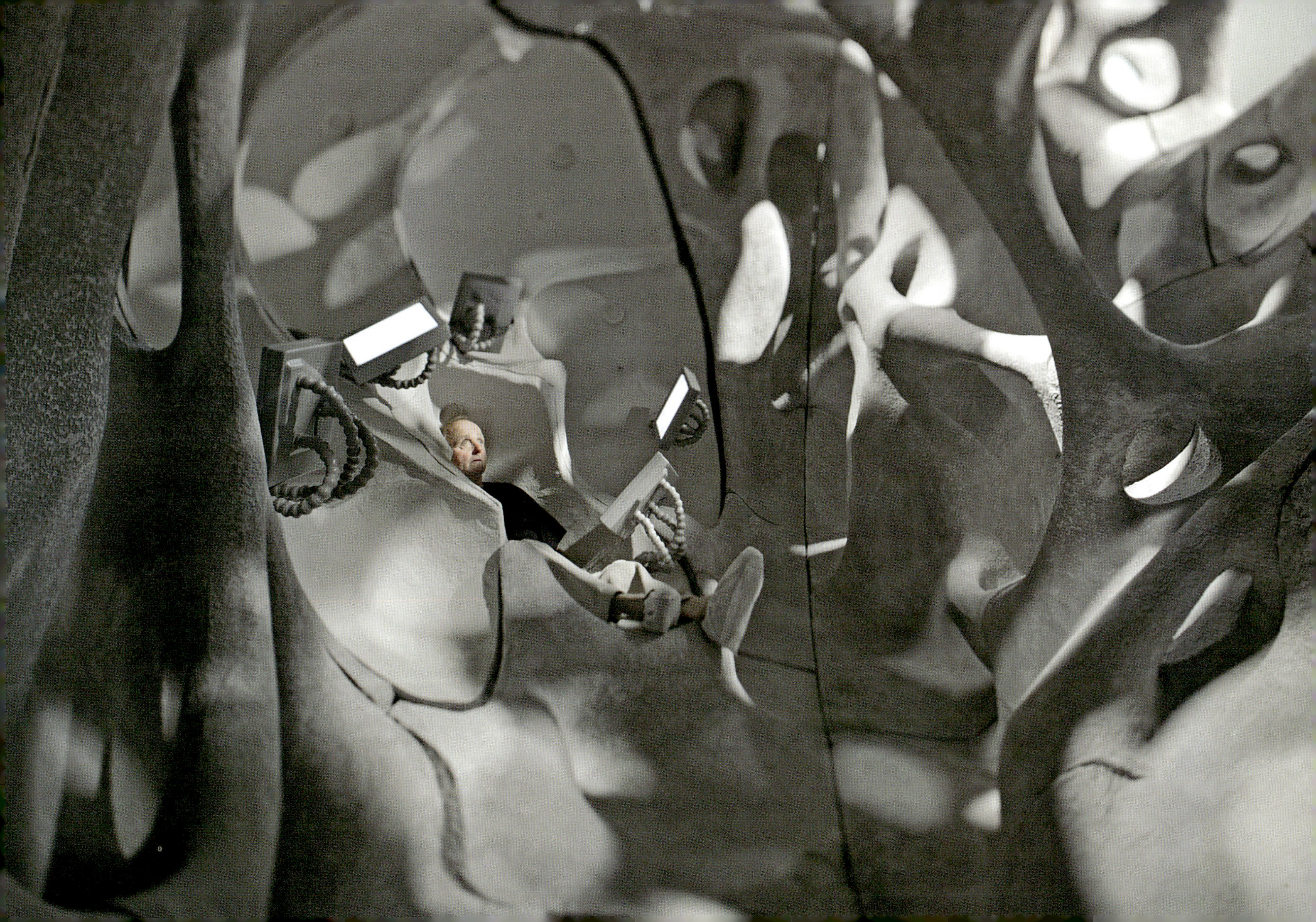

< multiple identity < utopia

fear and loathing

The complex interaction between individual and collective desires, between generative algorithms and protocols of structural compatibility, gives rise in **I've Heard About** to an open series of possible configurations, variable in space and time, determined by changes in the basic parameters. This nonhierarchical organization produces a system or relationship comparable to a natural phenomenon, a form of "swarm intelligence" (the collective behavior of a self-organized, decentered system) with particular biomorphic consequences that can be perceived on different scales of observation. Seen up close, spongelike structures emerge, similar to the internal structure of our bones. Different human activities take place in the gaps. On a larger scale, we see polypoidal layers, like coral colonies, extending in varied directions of growth. The overall appearance, however, is like a big ant farm, a tangle of filaments capable of resolving issues of circulation, distribution of static loads, and supply of services: a large vertical structure, crossed and overlaid by the system of extruding fiber cement, which terminates in coiling (and rather threatening) jointed arms. The outcome of the computer simulations of **I've Heard About** varies with respect to the functional expression of the radiolarians, nautili, and other simple beings described by biologist D'Arcy Wentworth Thompson, whose natural character corresponds to logical, balanced, Platonic mathematics, comprehensible and reassuring.[6]

Alternatingly, R&Sie(n) seem to express the most disturbing aspects of their nature when drawing on situations from the living world that are capable of triggering phobic reactions: spiders (**Swarm Town**, **Spidernethewood**), serpents (**Green Gorgon**), mosquitoes (**Mosquito Bottleneck**), excrement (**Mi(Pi) Bar**) and putrifying materials (**Aqua Alta**), or fermentation (**He Shot Me Down**). All this is combined with a taste for the display of innards, the creation of intestinal morphologies, and an apprecation of hairy surfaces studded with protuberances and orifices, slime green and flesh pink, foamy and gelatinous consistencies.

It would be all too easy to dismiss this delirium tremens imagery as the spawn of disturbed minds, or as the mere application of a tactic of disorientation that has nevertheless proven to be absolutely functional in other spheres (in the works of George A. Romero and Damien Hirst, for example). Many R&Sie(n)'s projects walk the perilous, mutable line that separates fully architectural action from art installation, and much of their provocative, liberating quality comes precisely from this borderline attitude. So the splatter aesthetic is not there just to resolve a problem of representation, or of mediated perception, the kind we can experience for a couple of hours at the movies, or at an exhibition. It interacts with more complex, interrelated levels of meaning, linking itself to the functional program, the interpretation of local conditions and life modes. Fear and loathing, precisely because they access a preconscious dimension—that of the survival instinct—become powerful vectors of transformation, formulating design strategies that are just as paradoxical as they are interesting.

disorientation

Let's look, as an example, at one of the basic functions of architecture on different scales: to provide information about our position in space; in other words, to help us know where we are and to allow us to enter into a relationship of identification with a particular place. Now let's try thinking about the opposite situation, the anguish of feeling lost, but also the lure of the unknown. The projects by Roche and his partners often try to generate such a situation, exploiting its potential to stimulate attention, surprise, and acceleration of experience, in a sort of concentrated reinterpretation of the Situationist dérive.

In **Green Gorgon**, the competition project for the art museum of Lausanne, Switzerland, getting lost is seen as an opportunity for continuous reprogramming and real-time modification of the use of the facility, based on individual desires. Disorientation is pursued on multiple levels: by confusing artifice and nature (the enclosure is a hydroponic cultivation of "spontaneous" weeds); by the tangling of constructive elements, creating a labyrinthine, three-dimensional circulation scheme; and by scattering hatchways, secret passages, and hidden shortcuts, as in a haunted house.[7] A GPS device combined with an information system permits visitors to interact with the building and the collection, indicating their position in space and allowing for different, unexpected itineraries.

A similar electronic device acts as an interface between visitors and **Olzweg**, a design for the addition to the FRAC (Fonds regional d'art contemporain) of Orléans. This project applies conceptual and procedural questions developed in **I've Heard About** to a specific situation. An explicit reference to *holzwege*, or paths that wind off into the woods—indicated by Martin Heidegger as a metaphor of human existence—speaks of an organism in continuous mutation, like a forest. In this case it is a pile of recycled glass bars, a greenish outpour that attacks the existing building, like a gelatinous secretion inside of which the rare openings that can be penetrated literally swallow up the visitor. The bars accumulate over a span of several years and are continuously repositioned by a mechanical robot arm whose operation is based on a stochastic program. The result is a condition of indeterminacy, of permanent spatial mutation and morphological slippage, preventing the formation of both spatial and temporal reference points.

The fact that museum functions (like urban utopias) may be plausible settings for this sort of architectural experience, similar to the simulated reality of theme parks, does not seem to be particularly important for R&Sie(n), who also come up with disorientation tactics for very different situations. The complex configuration of the **Loophole** footbridge—created to span the river that separates Cieszyn, in Poland, from Český Těšín, in the Czech Republic—is based, for example, on the interpretation of the place, its physical and political conditions, and above all, the experience of migration. Two opposite flows (going and coming, immigrants and emigrants, Poles and Czechs) are knotted and confused in the space, giving rise to a Borgesian "Garden of Forking Paths," in which the "right" way (the one that effectively makes it possible to reach the opposite bank) requires a detour from what seems to be the main route but is, instead, a dead end.[8] In this way, the layering of specific memories and sensations is brought to the fore, a result that would have been hard to achieve by means of simple, direct connection of the banks.

architects: R&Sie(n)... / Philippe Parreno   year: 2005   location: Lausanne, Switzerland   latitude: 46°30'N   average temperature: 49.5°F (9.7°C)

heterotopic-photosynthetic device, losing its identity, suffocating, sinking in its own green, ecological consciousness

client: Nouveau Musée des Beaux-Arts

a system of uncertain places

< green gorgon >

Map of botanical sequences:
The museum is an illusory space, more a landscape than an urban plan, more a forest than a work of architecture.
It begins to exist where different natures meet: the spontaneous grasses that come from the lake, an amphibious world on the way to emancipation; the urban nature of alignments, squares, and gardens, compositions that tend to dominate living organisms; the artificial nature of the green skin; the "enchanted" nature of spells, connected with our fears, with the childish wonders that bring gardens to life.

scenario

1)

Recognition of the "natural" artificial nature of the place (for example, the polder on the lake)

2)

Development of a heterotopic, tentacular, uncertain, organic structure

3)

Creation of scenarios of confusion between various environments—for example, man-made (hydro/aeroponic facade, biodynamic green hairs), urban, spontaneous, and haunted—in order to generate a hybrid, non-identifiable landscape

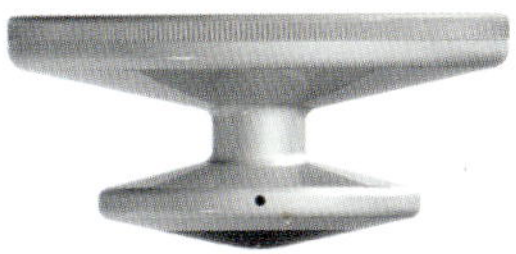

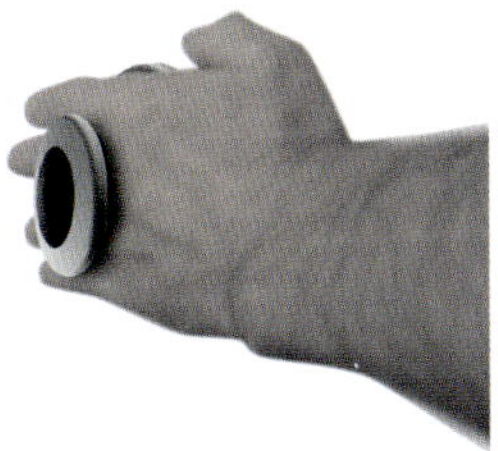

Views and screenshot of the sPIB device: A GPS navigator and vector of multimedia content, the sPIB provides visitors with indications of their position and audio-video information on the exhibited works. The sPIB can organize itineraries based on the preferences of visitors or, through a random search engine, interact in unexpected ways with the characteristics of the edified structure and its museum installations, proposing a nonpanoptical space.

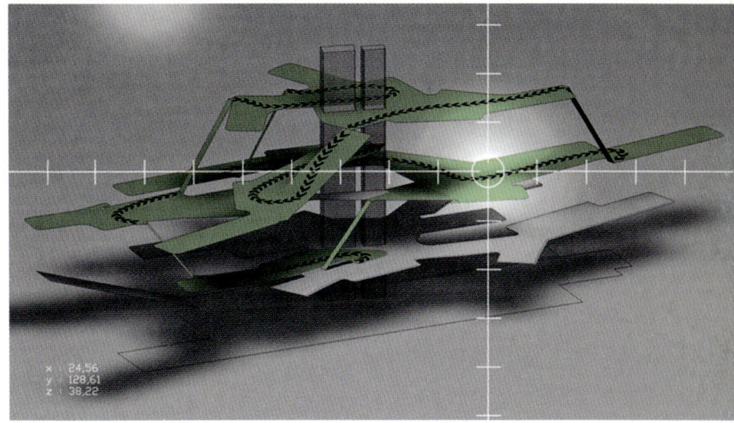

4)

Introduction of this knotty geometry into the interior morphology of the exhibition rooms: The result is a place where visitors can "lose themselves." They experience the collection in a process of "confused ambulation." Complexity is a tool of reprogramming and deprogramming, folding and unfolding.

5)

Individual positioning through portable GPS, linked to an informational pad: The visitor uses this i-compass both to move about and to obtain more information (sound and video) about each work of art on display.

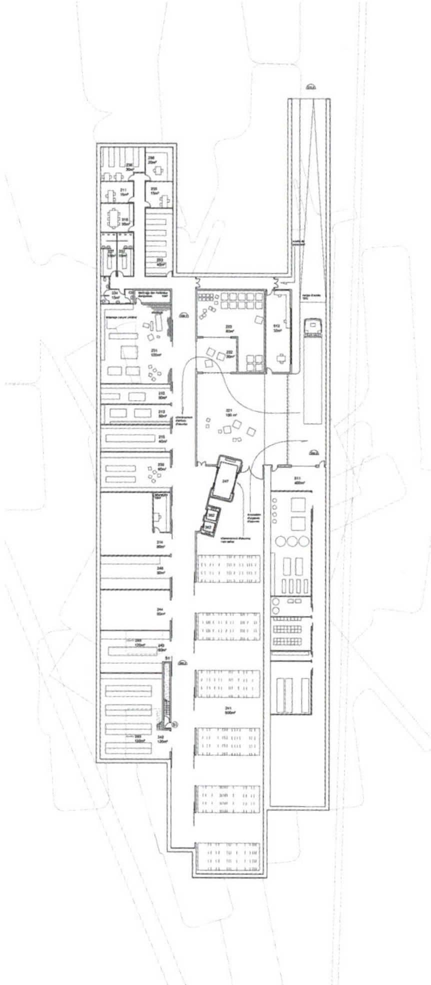

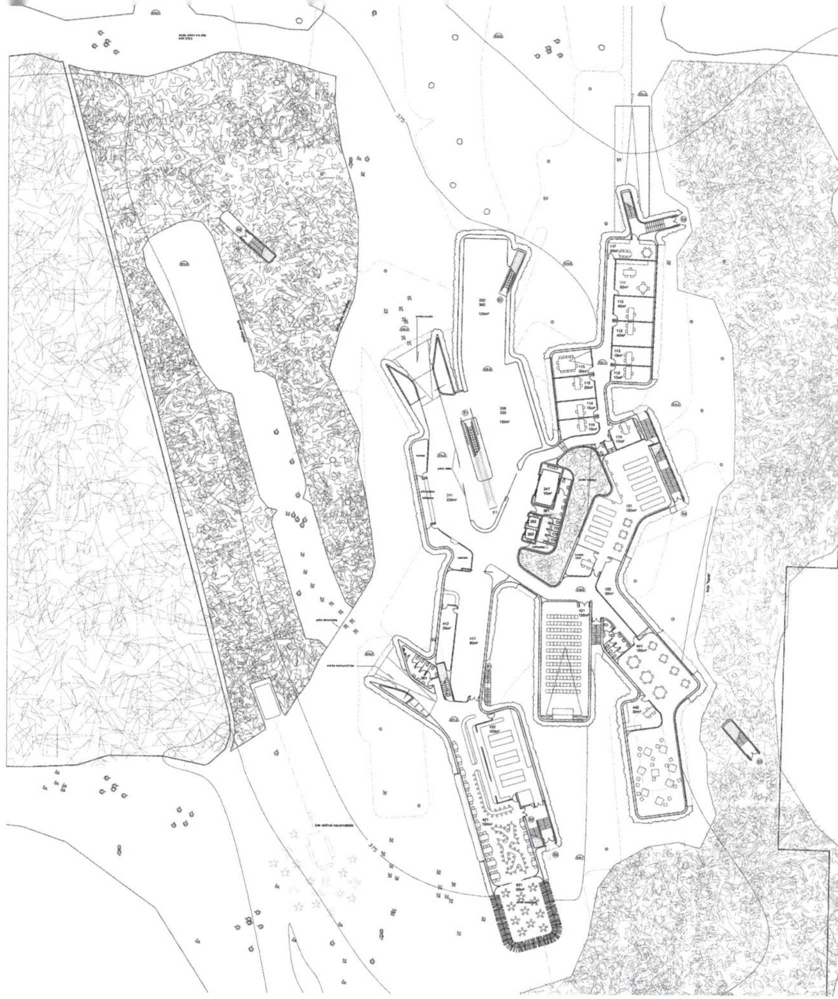

Basement and ground-floor plans:
The project for the Nouveau Musée des Beaux-Arts is a strange, enchanted dream that generates the extension of a lacustrian biotope—the undulating reflection of Ophelia's hair, the tangle of the forest in a fable from the Brothers Grimm: something fearsome that attracts us at the same time.

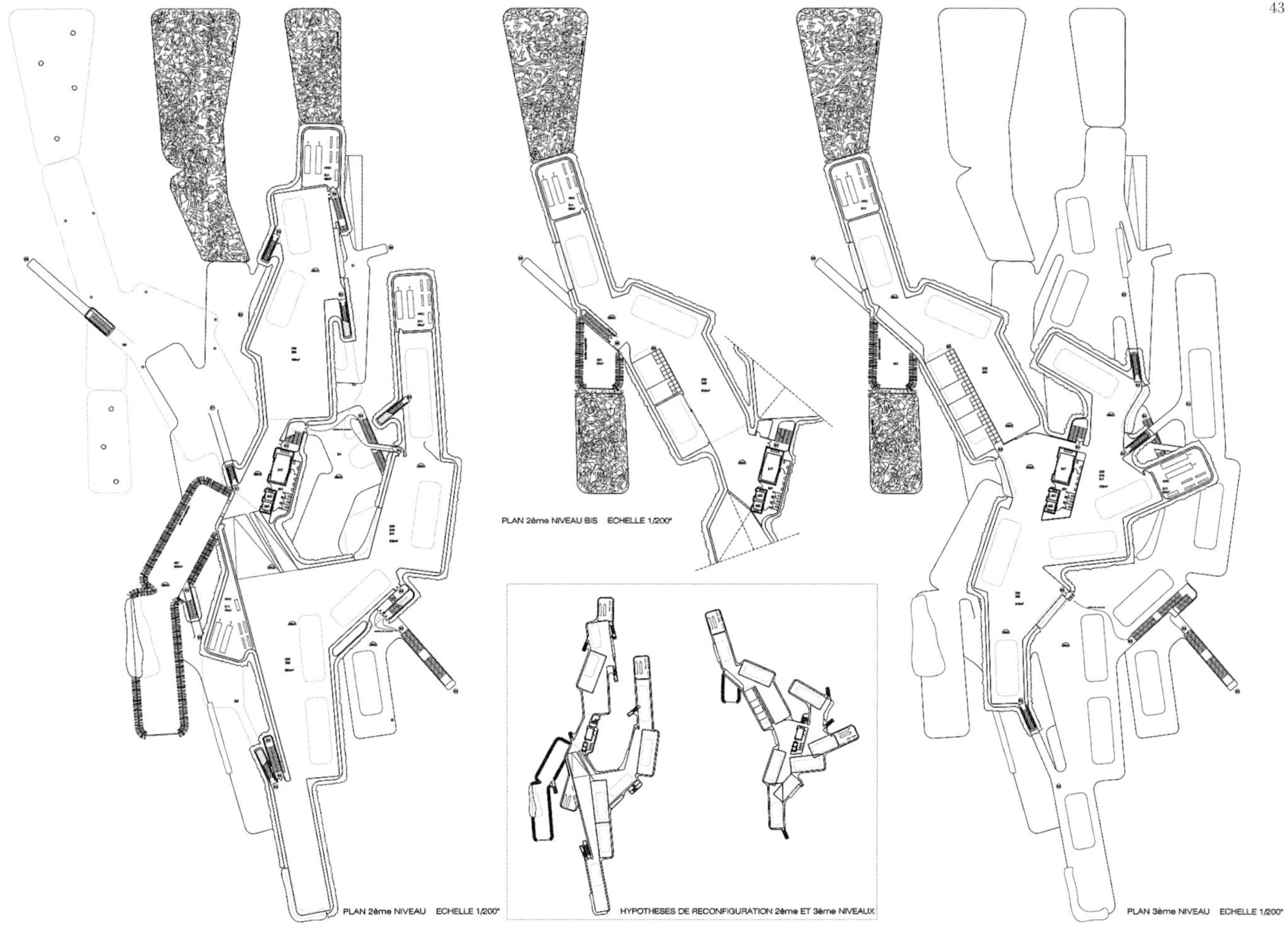

PLAN 2ème NIVEAU ECHELLE 1/200°

PLAN 2ème NIVEAU BIS ECHELLE 1/200°

HYPOTHESES DE RECONFIGURATION 2ème ET 3ème NIVEAUX

PLAN 3ème NIVEAU ECHELLE 1/200°

First-, second-, and third-floor plans:
The geometry interacts with the diversity of the collection and permits many installation hypotheses. Its labyrinthine conception in three dimensions responds to the various needs of the museum, making the required connections and separations. Different devices create temporary short circuits, crooked worlds, and fluctuating connections.

The museum multiplies its itineraries and dilates their timing. As in a haunted house, walls can be crossed, secret staircases appear, and hidden doors reveal forgotten rooms. Visitors can get lost or, more precisely, play with losing their way, using the sPIB as a tool to redefine their behavior inside a space designed as a Foucaultian heterotopia.

Entre tous les bocages du petit Parc de Versailles, celui qu'on nomme le Labyrinthe, est tout à fait recommandable par la nouveauté du dess(e)in.... Il est nommé Labyrinthe, parce qu'il s'y trouve une infinité de petites allées tellement mêlées les unes aux autres, qu'il est presque impossible de ne pas s'y égarer. Mais aussi afin que ceux qui s'y perdent, puissent se perdre agréablement, il n'y a point de détour qui ne présente plusieurs œuvres en même temps à la vue, en sorte qu'à chaque pas on est surpris par quelques nouveaux objets.

Charles Perrault, 1770

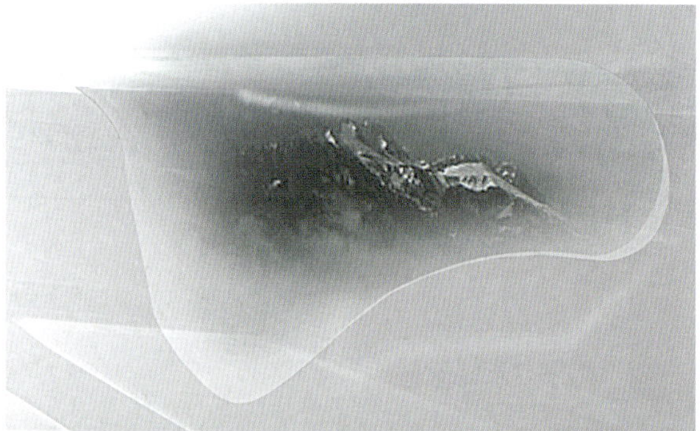

The landscape seen through one of the "eyes," made of ETFE plastic, and the same opening seen from the outside

Hydro/aeroponic devices on the enclosure and evolutionary phases of the botanical skin:
This biodynamic skin, besides permitting a confusion with the natural environment, filters urban dust and purifies the air.

architects: R&Sie(n)… /Pierre Huyghe   year: 2006   latitude: 47°54'N   average temperature: 51.6°F (10.9°C)

stochastic machine producing a repeating, deleuzian "body without organs"—an experiment with virtual potentials

client: competition, Fonds Regional d'Art Contemporain (FRAC)   location: Orléans, France

labyrinth of recycled glass

< olzweg >

Ground-floor plan and access system diagram:
The open conception of the project, connected with the "movement in the process of becoming" with which Deleuze defines the baroque (in *The Fold: Leibniz and the Baroque*), is particularly close to the collection and the policies of the FRAC: an adventure, a work in progress, just the opposite of the self-sufficiency of the sculptural, signage-type building.

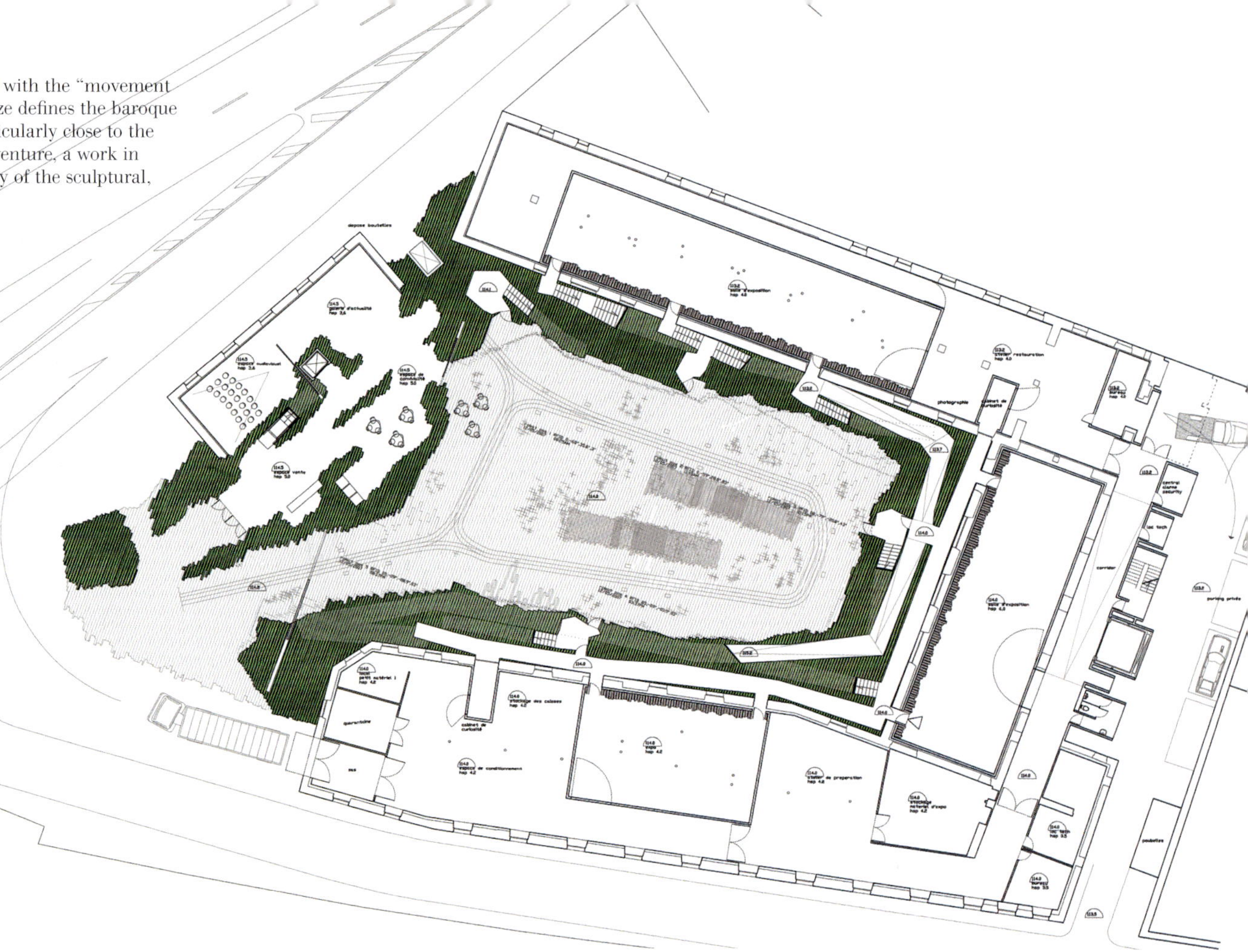

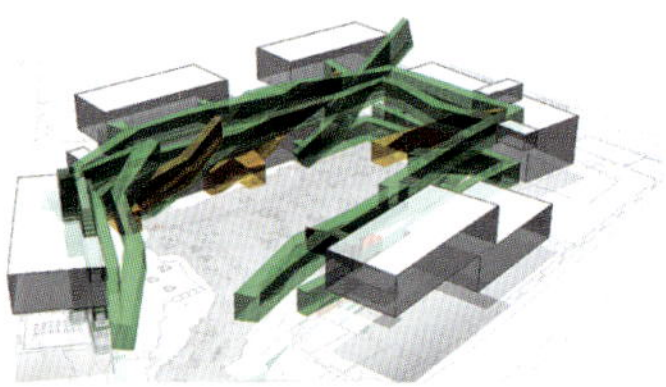

Olzweg takes its name from *Holzwege*, the book by Heidegger inspired by the paths that wind through the forest. A robotized construction system and a randomized, interactive software program permit continuous reprogramming of the parameters during construction.

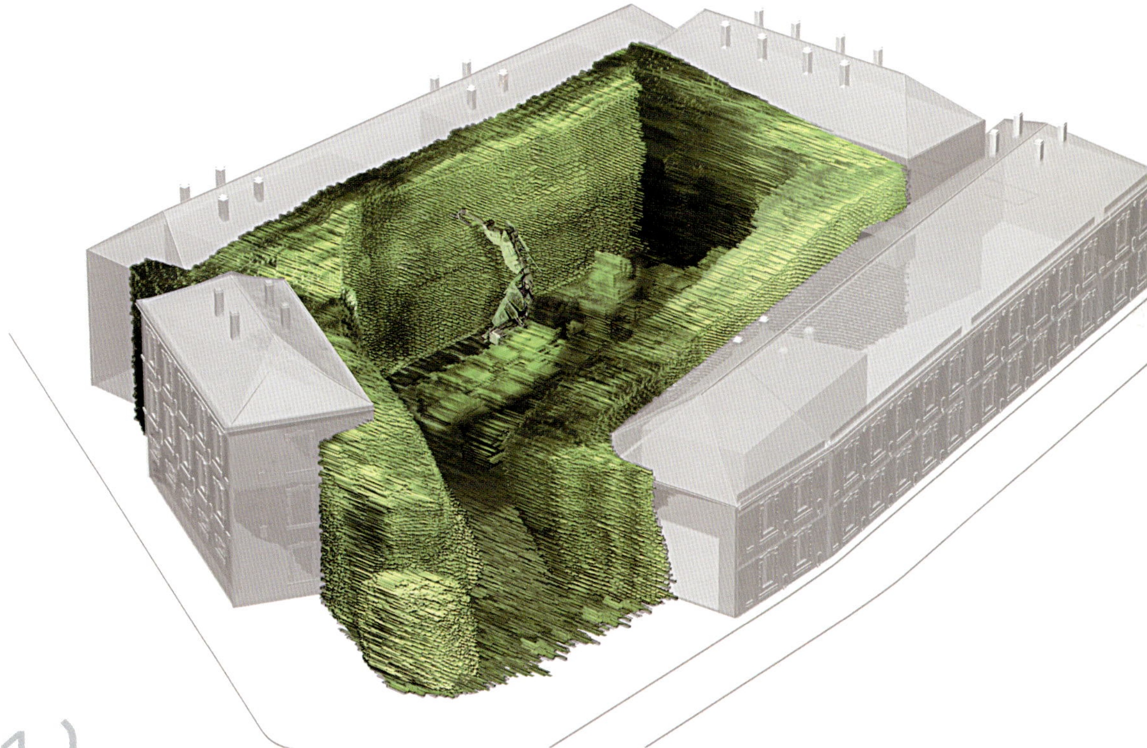

scenario 1)

The FRAC courtyard features glass sticks, which appear to be "glued" to the existing building (creating a "going to be done," incomplete process of construction). The glass sticks provide both a mazelike walkway and accessibility. A scattering script is written to develop this aggregation.

Virtual model of the robotic arm:
The robot takes part in the history of the project, in its literal, machinist, factual, and narrative inscription. It constructs both its real and its imaginary landscape.

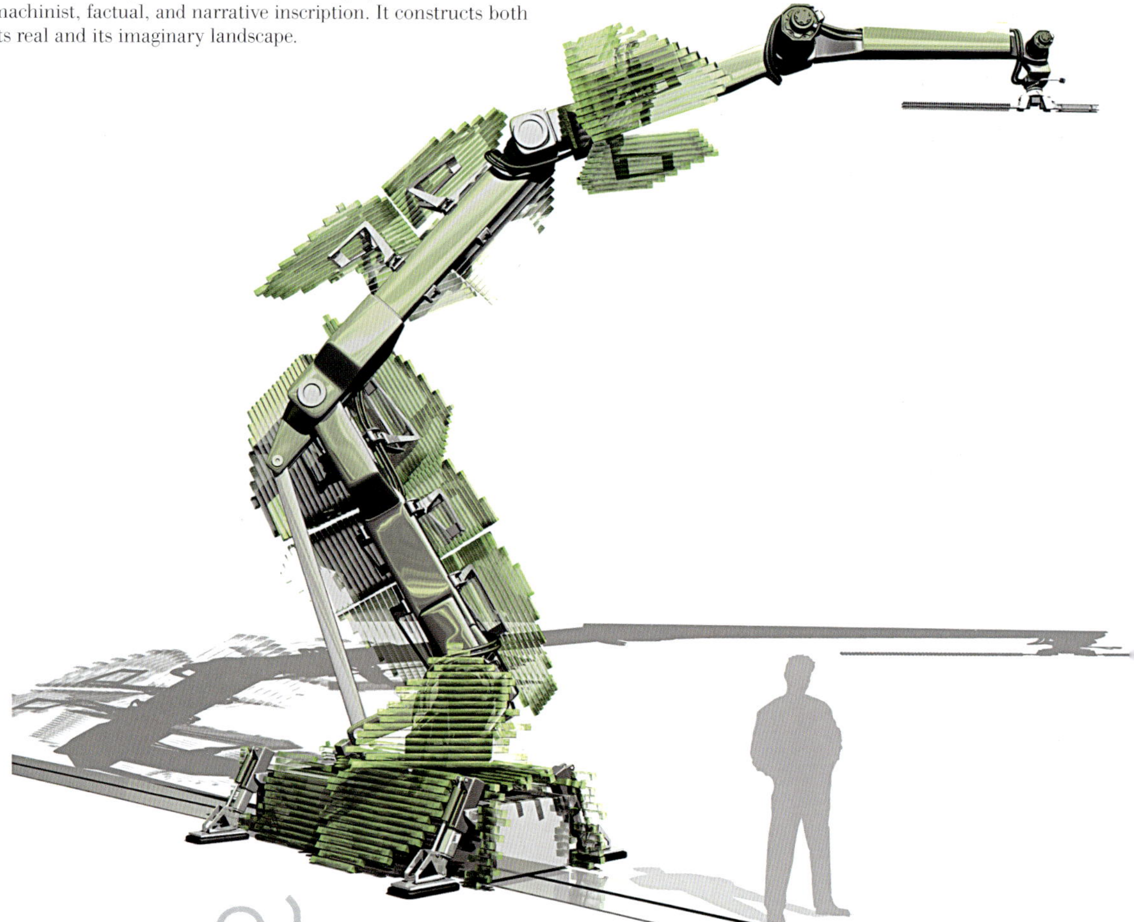

View of an interior:
The region of Orléans is one of the most active in the recycling of glass, processing twenty thousand tons per year. The museum constructs its own history based on a narrative of exchange with the place, not through the production of a promotional totem.

2)

Robots assist the construction and cleaning processes (both automatic and nonautomatic). This introduces randomness and uncertainty into the final shape. It also reprograms the labyrinth during the construction process.

3)

The glass is provided by a neighborhood recycling scheme. This helps reduce costs, while at the same time involving the inhabitants in the "work to be done" experience.

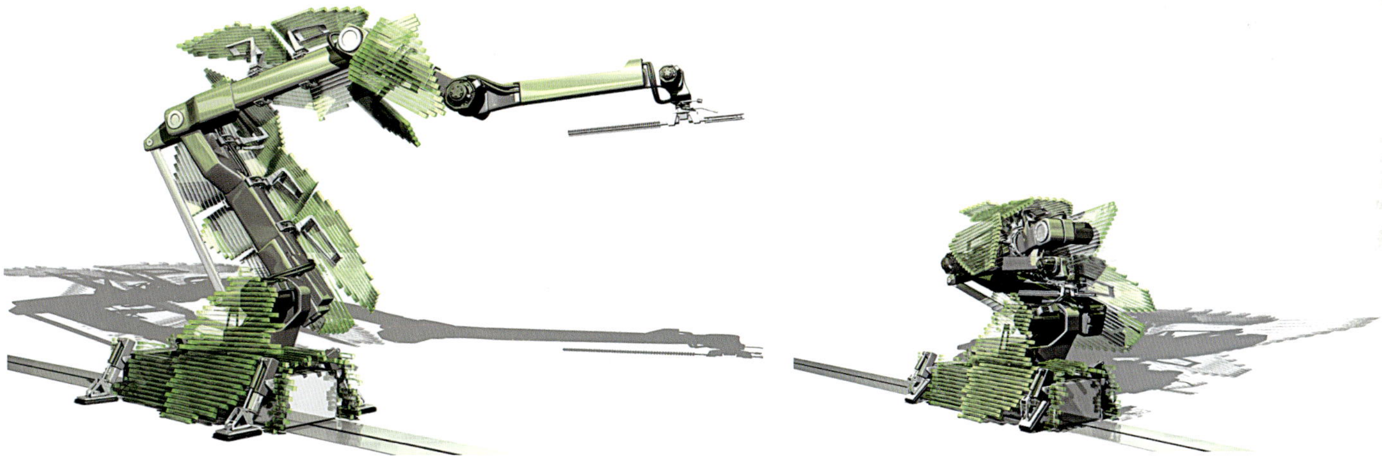

4)

For the same reason, the construction schedule stretches over a ten-year period: "Construction becomes a factor, a desire with a desirable machine" (Duchamp).

The courtyard, with the robotic arm, and one of the internal routes:
The robot establishes a symbiotic relationship with the glass, like that between a shark and a pilot fish.

5)

The use of an RFID pad obliges visitors to accept the fact that they are lost in the glass maze. They can also redefine their XYZ positioning in relation to the specifics of the exhibition (PAD with sound, video, and local GPS).

architects: R&Sie(n) year: 2006–8 latitude: 49°45'N average temperature: 46.8°F (8.2°C)

identities confused by forking paths that articulate both emigrant and immigrant statuses

client: city admininstrations of Cieszyn and Český Těšín location: Cieszyn, Poland / Český Těšín, Czech Republic cost: $2,624,000 (€2,000,000)

bridge on the Czech-Polish border

< loophole >

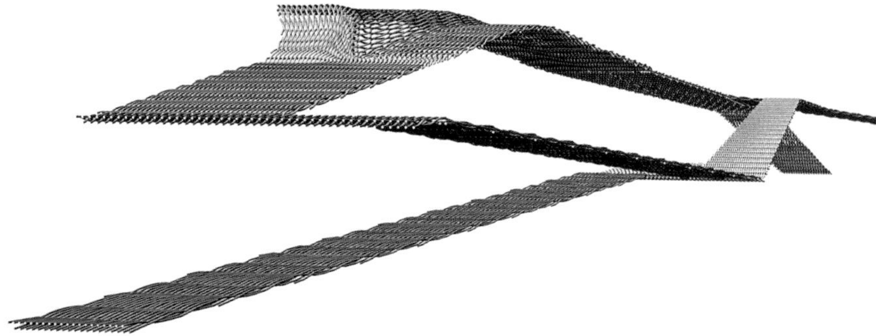

scenario 1)

Recognition of the area's local history and its natural and political difficulties: to cross the border over the river

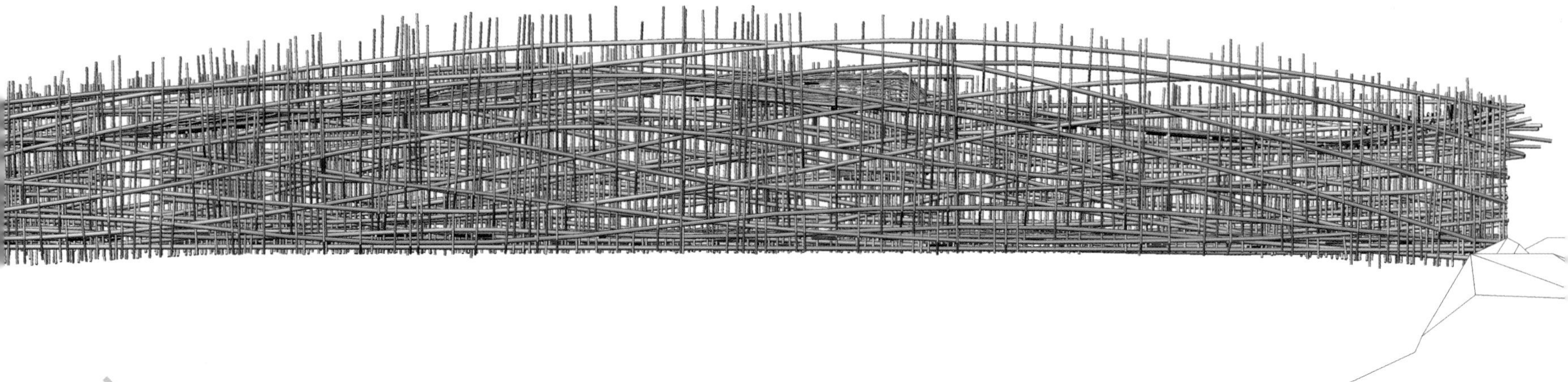

2)

Development of a strategy to blur the distinction between "emigrant" and "immigrant"

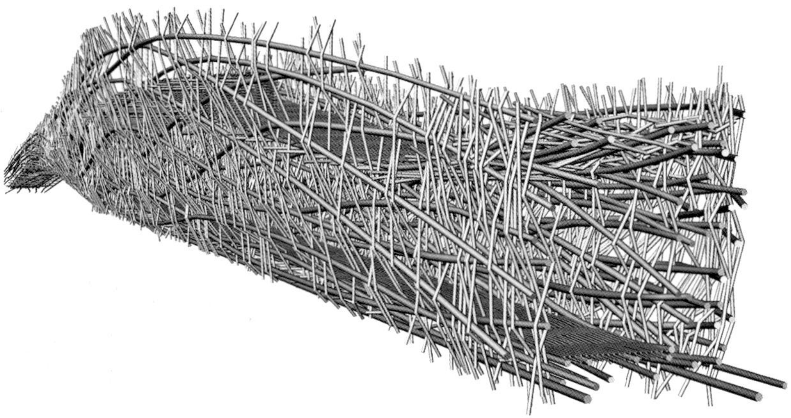

3)

Production of a loop and a tricky walkway, to make the act of crossing ambiguous

4)

A straight line is never the best way to join two points in space.

Structural solutions:
Random vertical uprights are woven with horizontal structures (like a nest) in pairs, using "clips," developed with architect Marc Fornes.

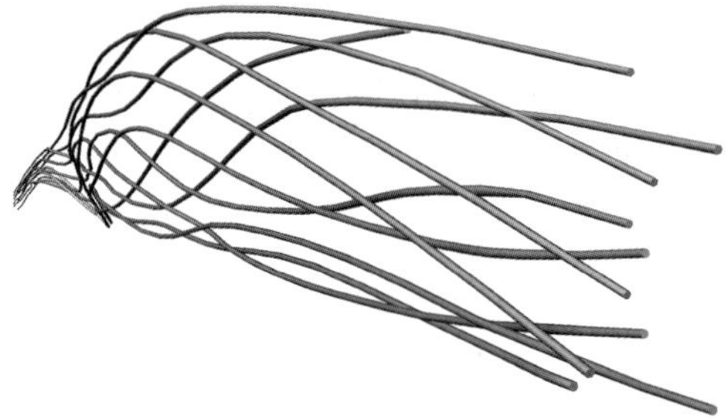

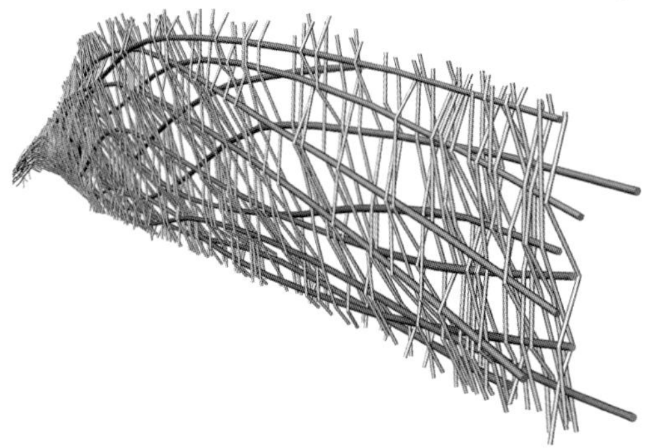

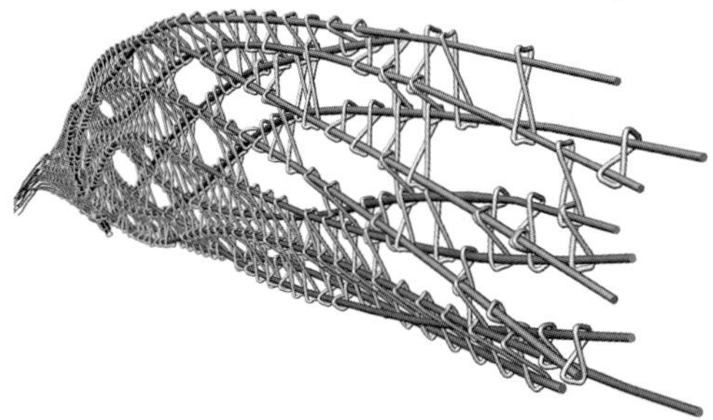

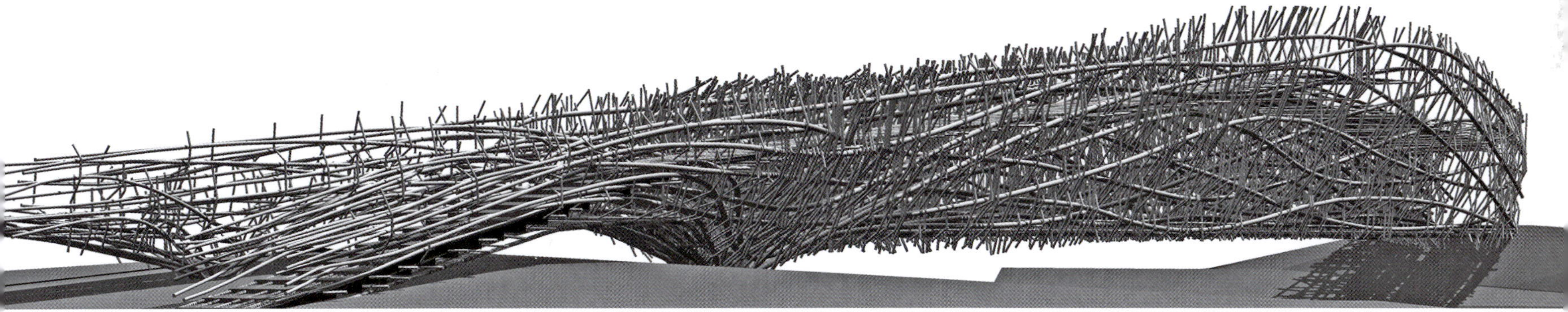

A "crafted" solution:
The bridge absorbs the lack of symmetry between the Polish bank (right), protected with an artificial embankment, and the Czech side, where the deposits form a more gentle slope.

The labyrinthine appearance of the bridge is accentuated by the random, confused quality of its structural weave: like the wavy hair of John Everett Millais's *Ophelia*, floating in the water.

The organic connection of the structural partitions helps to materialize a definite space: the womb of a living body.

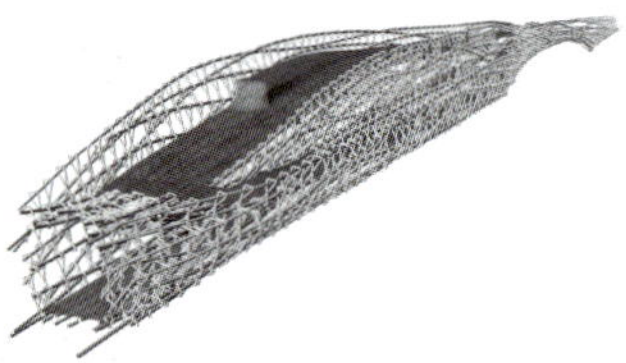

Parametric solution:
The bridge is a territorial fold on the border, a Borgesian "Garden of Forking Paths": to cross it one must turn back and discover a disguised passage, hesitating between the condition of an emigrant and that of an immigrant.

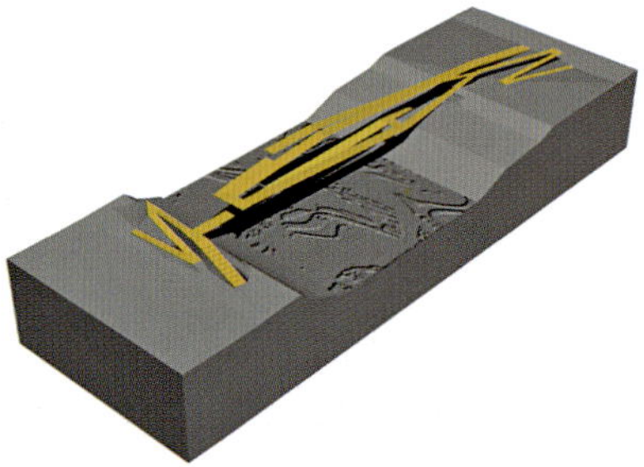

< multiple identity < utopia < fear and loathing < disorientation

furtive tactics

The techniques of detour and disorientation connect, in other projects, with the paranoid interpretation of local conditions. The Barak house (**Shearing**), built in southern France, overcame the impossibility of building on its lot due to landscape restrictions protecting the view of a nearby medieval tower. The solution can be defined as mimetic, in both visual and conceptual terms. Ripples on the ground—made with green plastic and authorized as a technique of agricultural protection—create the physical and regulatory space for the construction of a "stealth" building, invisible to the "radar" of the environmental protection authorities. The actual house, built afterwards inside this authorized structure, escaped being sentenced to demolition. The payment of a fine, already included in the overall cost estimate, sufficed to make the project completely legal.

Another such furtive tactic can be seen in **Spidernethewood**, a large vacation home, also in the south of France. Here the vegetation, thickened by the introduction of new plants, is separated and restrained by nets that define a series of spaces and labyrinthine paths. The building proper, a simple parallelepiped, becomes indistinguishable from the tangle of trees and bushes, and like them it is crossed by a system of voids. The result is a sort of continuous "public" space, ambiguously suspended between interior and exterior, whose only visible transformation is the densification, near the building, of the weave of the fabric that encloses it.

A more immediate, concrete threat is perceived in the project for a multifunctional complex not far from the demilitarized zone between the two Koreas (**He Shot Me Down**). In this case, fear represents a specific factor of local identity, and as such it is inserted into the design strategy. The building is positioned as an extension of a mountain, like a bunker, once again concealing the new layout. Only a series of ballistically inspired openings, like holes made by giant bullets, reveal its interior of fluid, gleaming metal: a reference to the T-1000 in *Terminator 2*, the hypermimetic android assassin capable of imitating any living being simply by touching it, whose metal insides were revealed only through dramatic extrusions and lacerations.

architects: R&Sie(n)... year: 2001 location: Sommières, France latitude: 43°50′N average temperature: 57.4°F (14.1°C)

a delicate and impermanent architecture waiting, as in the three little pigs fairy tale, for the wolf's blow...

client: Judith and Ami Barak cost: $160,000 (€121,900) dimensions: 1,720 ft.^{2} (160 m^2)

the invisible house

< shearing >

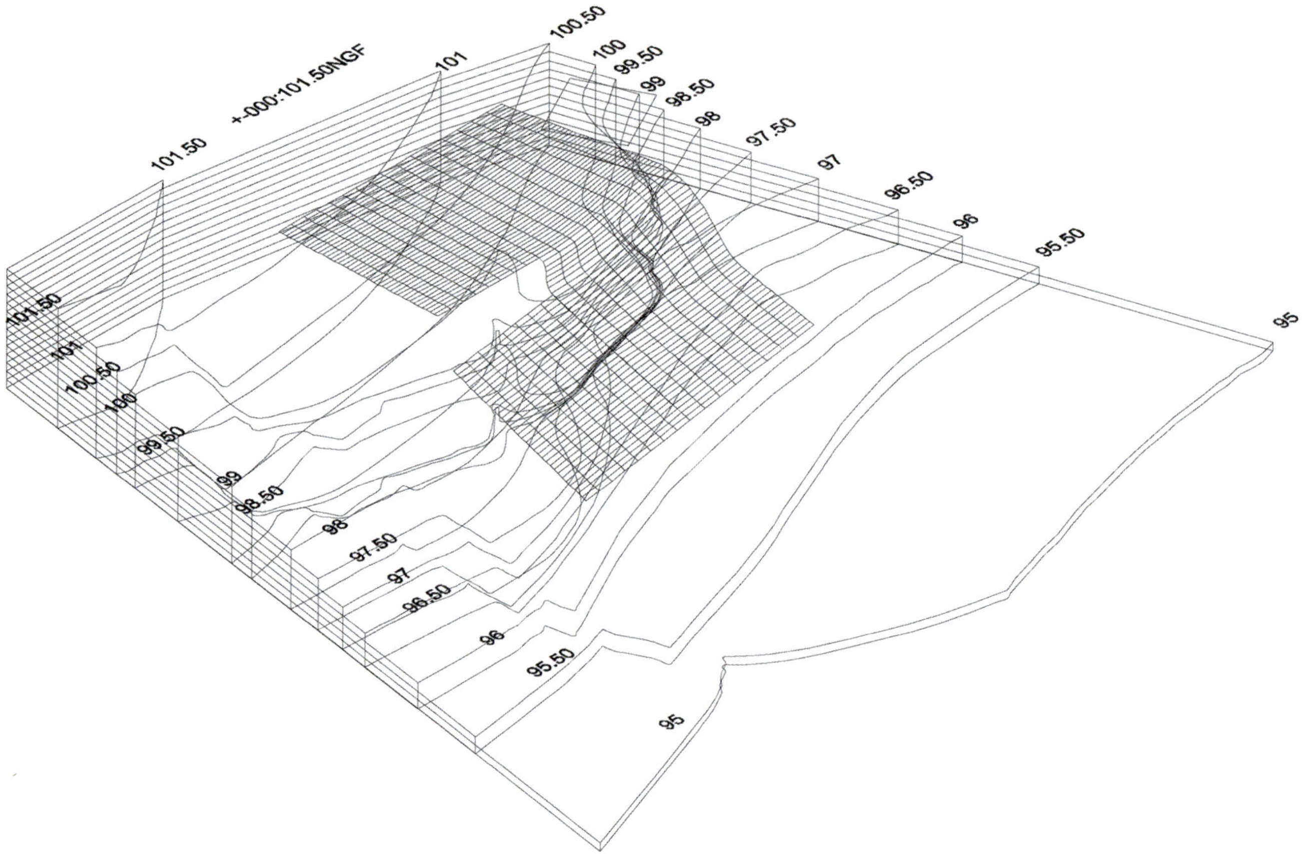

The house in the landscape and a virtual model of the enclosure on the topographical survey of the terrain:
The area is subject to landscape conservation regulations due to the presence of a nearby medieval castle. The terraced lot, which is difficult to reach, is divided into two parts by a support wall, with a level shift of over thirteen feet (four meters).

scenario

1)

Deterioration of the landscape, like a new geological fold, to obtain stealth

2)

Design of a house—like a sheared, upthrust rock layer—alongside an existing stone wall in the middle of a field

The interstitial spaces between building and enclosure (awaiting colonization) and the entrance facade:
The house uses two construction systems: the skin, made with resilient plastic tarps usually used in agriculture and stretched on a metal structure, was assembled by a company that specializes in art installations; the house proper was built by a local contractor using traditional masonry techniques with a stucco finish.

3)

Construction of a tentlike space to provide protection from the weather, with living areas inside

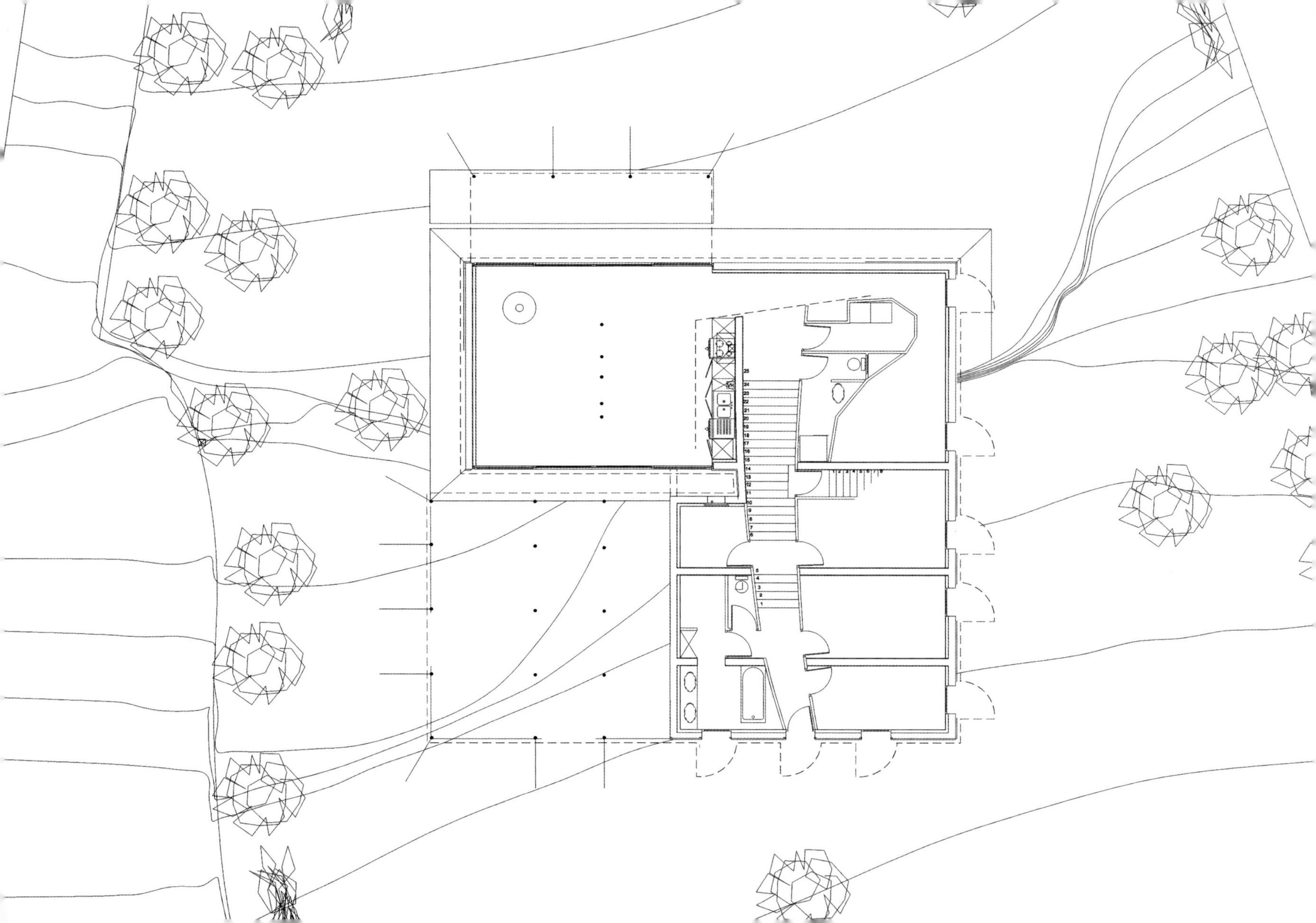

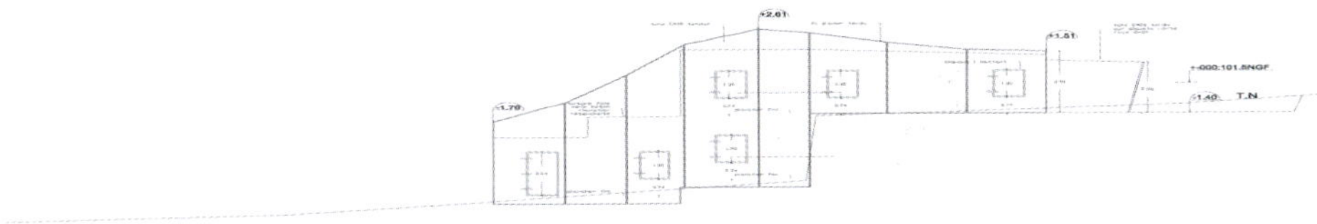

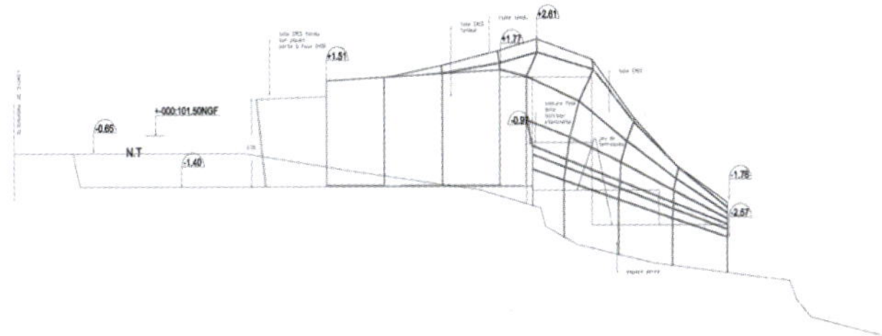

Elevations and detail of the enclosure

General plan:
The client, the director of a contemporary art center, wanted a seven-room house for a family of four.

The living area, kitchen, and internal circulation layout

architects: R&Sie(n) year: 2007 location: Nîmes, France latitude: 43°50'N average temperature: 57.4°F (14.1°C)

fragile labyrinth for stealthy and furtive occupation

client: Urbain and Elisabeth Souriau cost: $918,700 (€700,000) dimensions: 4,300 ft.^{2} (400 m^2) + outdoor labyrinth of 21,500 ft.^{2} (2,000 m^2)

labyrinth amidst branches, house without facades

< spidernethewood >

The “textile” labyrinth, inside and outside, and a view toward the park

scenario

1)

Over-densification of existing forest (the trees should reach the right level in five years)

2)

Netting and wrapping of the predicted size of adult trees with a polypropylene mesh: This develops a labyrinth in the branches.

3)

Inclusion of a stealthy, indoor, 4,300 ft.^{2} (400 m^2) summer building on two floors: This is connected to the labyrinth by a huge 23 x 11 ft. (7 x 3.5 m) sliding glass door.

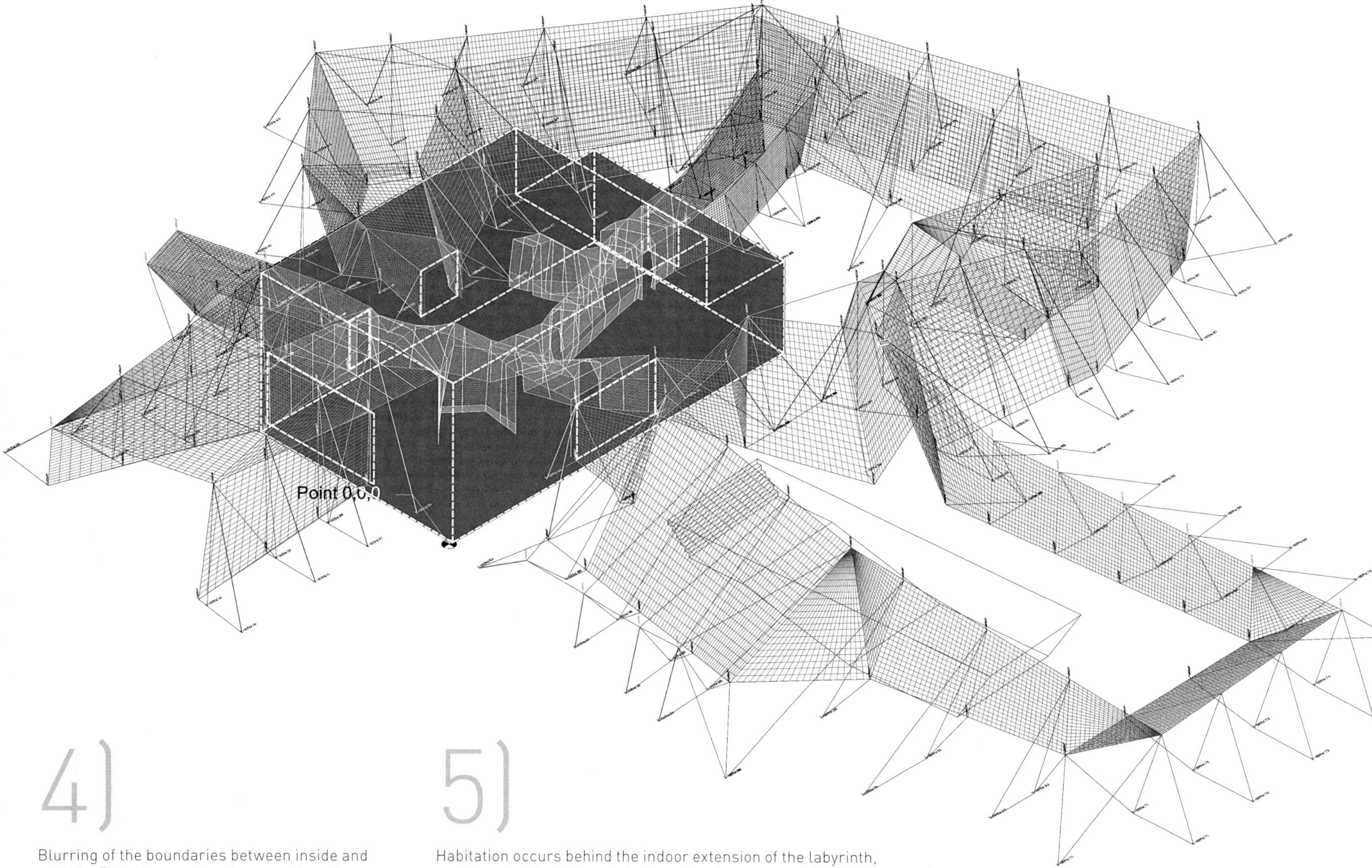

4)

Blurring of the boundaries between inside and outside: This also creates the sensation of a refreshing wind.

5)

Habitation occurs behind the indoor extension of the labyrinth, past the plastic-strip curtain, in an "under-construction" stage, with a standard design from the corner-shop mall.

Virtual model and views
of the "textile" labyrinth

6)

A "lost game" in the neighborhood (in five years) for architecture without any facade

General plan and interior views

architects: R&Sie(n)... year: 2006–7 location: Heyri, Korea latitude: 37°47'N average temperature: 52.7°F (11.5°C)

recognition of the eroticism resulting from human conflicts and biomass collecting machines for a haunted house

client: Miss Julieta and Mister Lee, Seoul cost: \$3,000,000 (€2,286,000) dimensions: 21,500 ft.^{2} (2,000 m^2)

a robot on tracks

< he shot me down >

going into the forest

The robot for collection of biomass, and view of the building:
The concrete threat of weapons and minefields, from which humans are barred, has made the demilitarized zone a natural oasis. This absence represents the fundamental theme of the project, whose open spaces are crossed only by the biomorphic robot used for the collection and scattering of grass, leaves, branches, and other items from the forest.

scenario 1)

Recognition of the situation in the demilitarized zone on the North Korea–South Korea border: Here danger is woven with paranoia and its theatricality.

2)

Recognition of the specificities of the urbanism of Heyri, where domesticity is fully embedded in the public area and the client becomes the support for public services

3)

The sense of paranoia becomes erotic, thanks to the sweet ballistic impacts in a volume extended from the mountain that absorbs the program.

4)

Creation of bullet trajectories as walkway access and light providers, made of stainless steel and glass as an homage to T-1000

5)

Development of a robot that runs on a specific track into the forest to bring back biomass on its back: It smears the decomposing biograss and bioleaves onto all the external surfaces of the building. This helps insulate the building.

Ground- and first-floor plans and sections:
A multifunctional building with private home, disco, shop,
restaurant, and children's museum

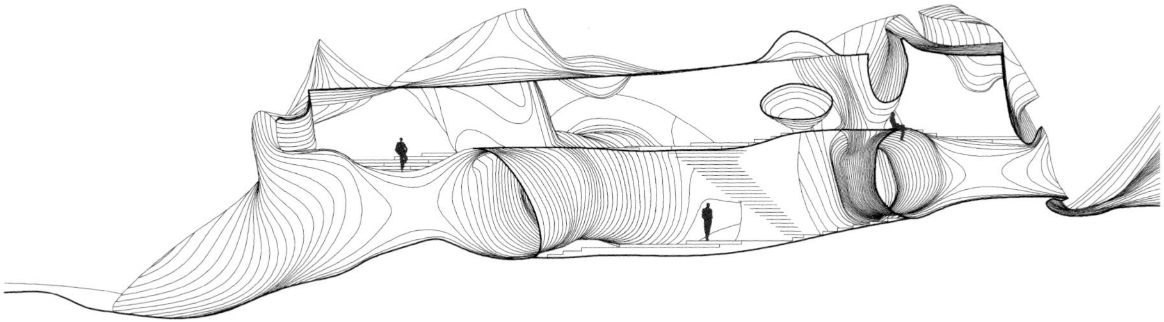

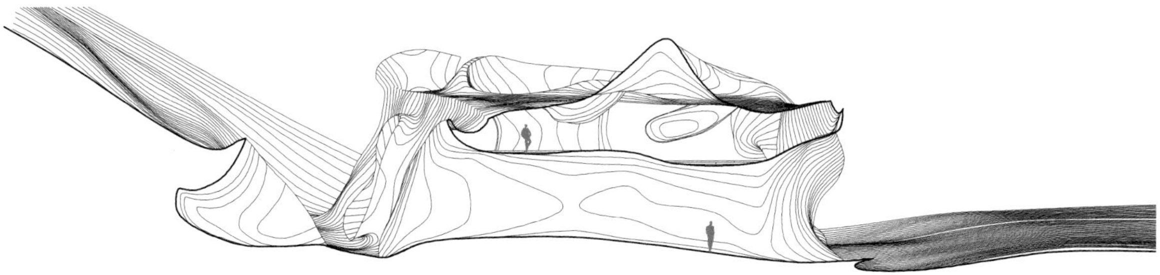

Views of the demilitarized zone, position of the project, and views of the model

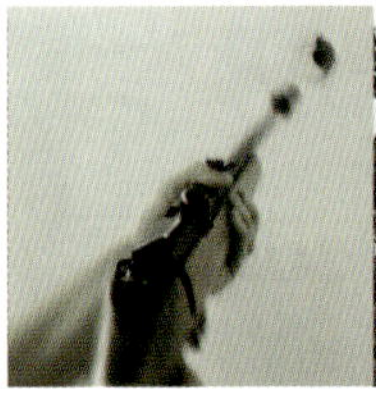

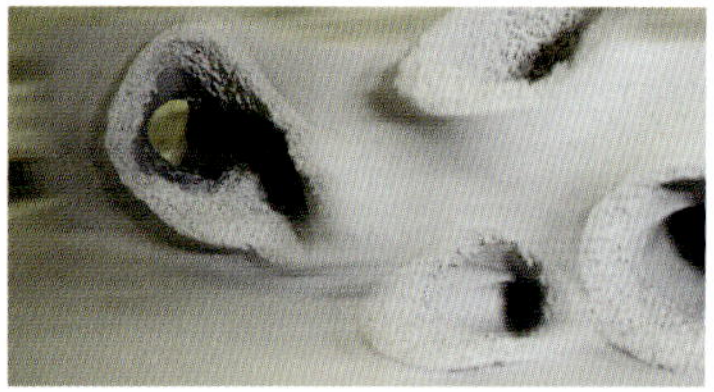

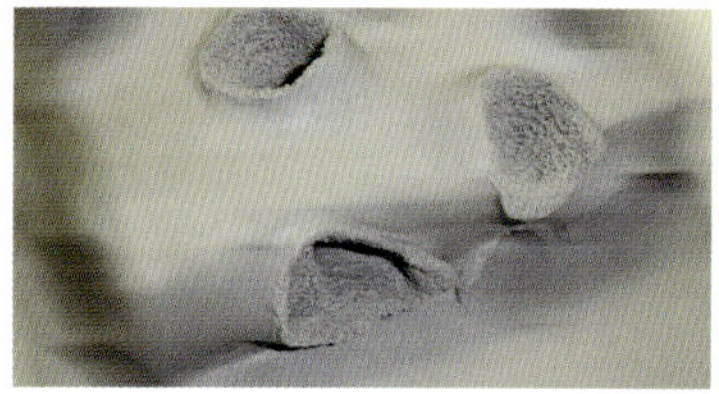

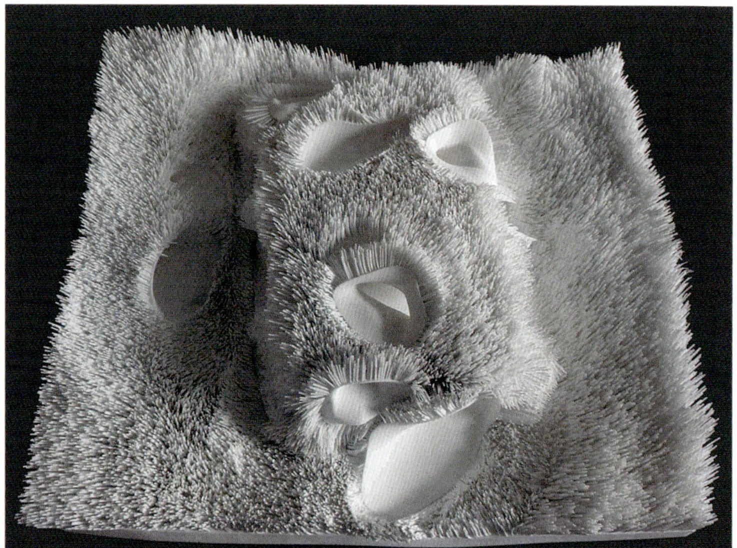

< multiple identity < utopia < fear and loathing < disorientation < furtive tactics

base materialism

The vast terrain in Korea is treated as inaccessible nature, deprived of human presence, as in a minefield or another place set aside for military devices located in the border zone. It is a place explored by insectoid robots sent to gather fallen leaves, grass, and other biomass, which will ferment, producing heat that contributes to the climate control of the complex. Along with their use of paranoia as both a motor of interpretation and a tool to shift our way of looking at things, here Roche and partners overlay one of the characteristic features of their recent work, namely, the attraction to decay, corruption, and "base materialism" (to use a category from Georges Bataille that is certainly a significant presence in their reflections).[9]

Depleted, entropic matter, expelled from the production chain or degraded by natural processes, becomes a surprising resource for conceptual activation. For example, in the **Mi(Pi) Bar**, an installation at MIT, visitors are supposed to drink their own urine. Beyond the various considerations regarding the lack of danger involved in this practice (as long as the urine is fresh, and personal) and its therapeutic benefits, this project by R&Sie(n) makes use of a series of stratified fears triggered by any relationship with bodily fluids. It does so by placing machines between these paranoias and their object, machines that can be regulated according to the needs of the visitor, who can program the percentage of urine filtration to the point of obtaining pure water flavored with tea. Besides our disgust for the fluid itself, possible suspicions also arise about the technology and its efficacy in guaranteeing the promised concentrations, and above all its prevention of any possible mixing of the "contributions" of different persons. The installation develops from an idea proposed for the Venice Biennale in 2000, **Aqua Alta**, where the guests of a boat could drink specially filtered canal water (whose composition is best left unreported). In Cambridge, though, the paranoid spiral of the operation is further accelerated by an inflatable enclosure that protrudes from a building by I. M. Pei: a yellow, foamy formation of disturbingly evocative natural appearance, produced by a generative script.

With respect to the stances criticized by Karl Kraus's aphorism on Loos regarding the urn and the chamber pot, we are on another planet.[10] The approach of R&Sie(n) moves nimbly beyond both the decorative obsession of those who dress up a bedpan as a funerary urn, and the functionalist rigor of those who want to make an object of great symbolic impact into something utilitarian, and end up designing an urn as if it were a chamber pot. Instead, what emerges is a direct reference to Duchamp's "readymades," especially *Fountain*, and the series of shifts caused by the idea of displaying a urinal as art, with its position (and function) inverted.[11]

A readymade by R&Sie(n) is stripped, however, of its quality as an object, the serial nature of its production, the reversibility of its use. It might be more precise, here, to talk about readyused or, even better, ready(ab)used, an instrument of radical acceptance of the world, in which every prejudice is inexorably overturned.

architects: R&Sie(n) year: 2007 latitude: 42°21'N average temperature: 49.5°F (9.7°C)

psychodigestive, paranoid negotiation with the illusion of toxicity

location: extension of I. M. Pei's Wiesner Building, MIT Campus, Cambridge (Mass.), USA cost: $300,000 (€228,600)

excrescent wrapper for self-consumption

< mi(pi) bar >

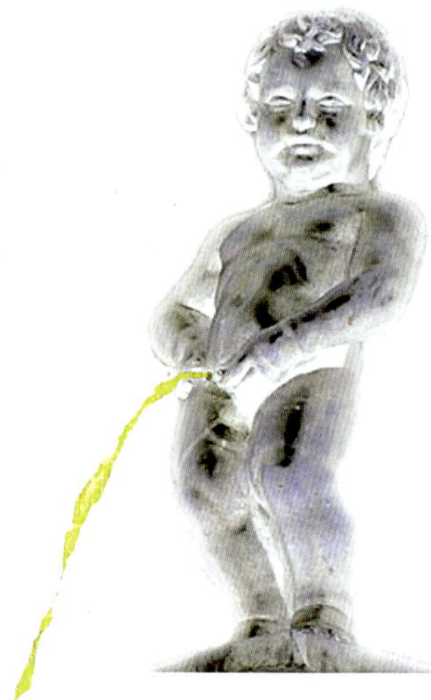

scenario

1)

Drinking one's own urine could in fact trigger a healing process: urine is a byproduct—rather than a residue or waste—of the body's blood filtration system.

2)

Exploitation of fears and anxieties about the toxicity of bodily fluids—their assumption and exchange

From the producer to the (self)consumer:
Close-up of the wrapper

3)

Formation of a filtering machine and a protocol that could avoid disgusting and "dangerous" mixtures

4)

Introduction of a suspicion about the effectiveness of technology

5)

Design—through a parametric script—of a scummy, yellow excrescence:
The morphological distortion of the tiled exterior of Pei's building accelerates this sense of paranoia.

The automatic dispenser:
The bar offers the possibility of generating an immunoactive substance through an infinite process of absorption, digestion, defecation, and reabsorption. The purpose is to provide the possibility of reintroducing bodily secretions as positive entities with which we must come to terms and negotiate.

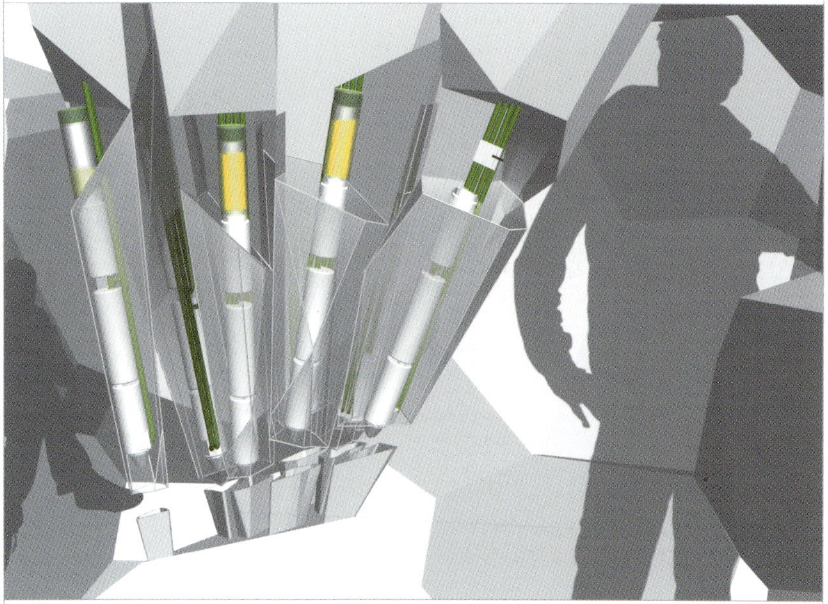

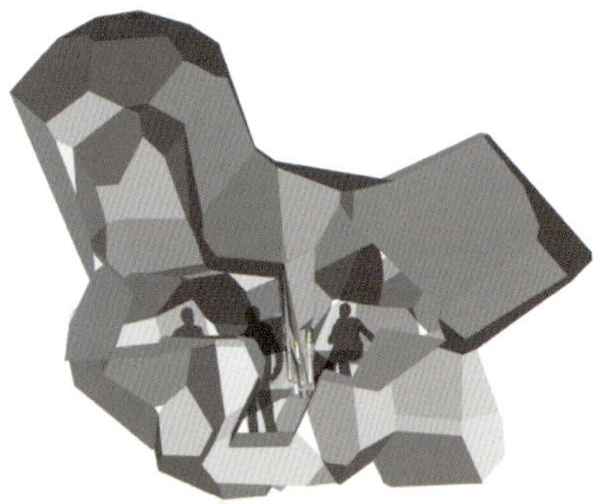

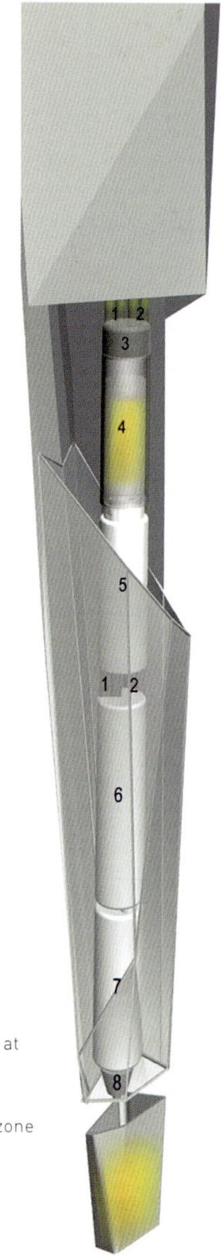

1 ozone conduit
2 heating conduit
3 cover
4 urine container with plug at bottom
5 filter
6 adjustable-percentage ozone filtering system
7 addition of tea / warming
8 dispenser

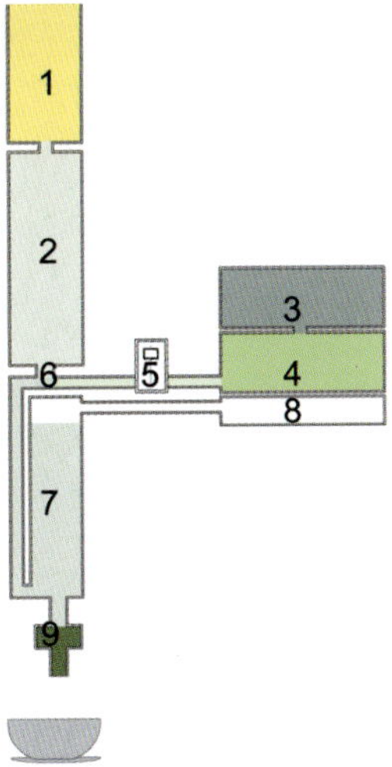

1 introduction of urine
2 filtering
3 oxygen canister
4 ozone generator
5 ozone control
6 Venturi injector
7 ozone recycling
9 addition of tea / warming

Diagram of the filtering system and positioning of the equipment in the installation

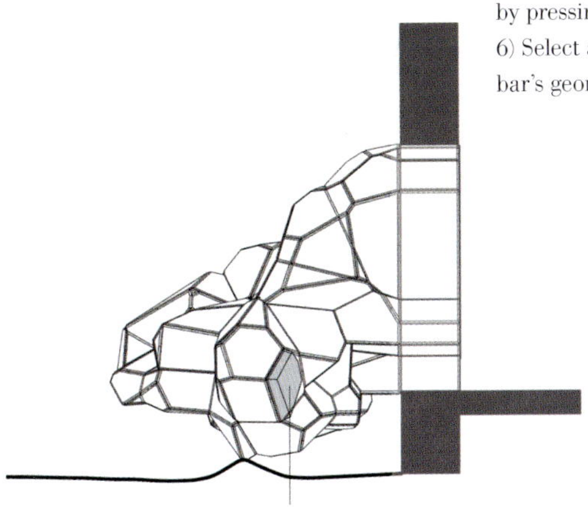

User manual:

1) Clean one of the seven tea infusers by pressing the button "Wash My Filter."
2) Use the adjustable dimmer switch to determine the desired degree of filtration from 0 percent to 100 percent.
3) Place the container in the cavity designed for this purpose (the central, stalactite-shaped area).
4) Remove an individual teacup and saucer, as well as a tea bag, from the vertical dispensers (self-service).
5) Start filtration and infusion of tea by pressing the button "Mi(Pi)."
6) Select a seat of your choice in the bar's geometric interior.

Mi(Pi) Bar, phases of automatic generation of the volume

< multiple identity < utopia < fear and loathing < disorientation < furtive tactics < base materialism

danger

One particularly clear example of a ready(ab)used is the project for a house on the island of Trinidad, located in a zone threatened by mosquitoes that carry the West Nile virus. **Mosquito Bottleneck** combines the threat of contagion (and the sensations of irritation and repulsion caused by the presence of insects) with devices made by the local population for defense, namely traps in which the upper part of a plastic bottle that has been cut in half is inserted upside-down in the bottom half: the remains of the soft drink the bottle once contained attract the insects, which, once inside, are unable to find their way out. The building, then, is a double enclosure significantly woven with very slender ribbons of transparent polyethylene made from bottles. Inside, the inhabitants can enjoy listening to the buzzing of dying mosquitoes trapped in the interspace, thanks to resonance within the structure. The fact that the house—shaped around the spiral trajectories of the insects—resembles a piece of digestive apparatus set down in the luxuriant natural setting of the Caribbean simply intensifies the peculiar sense of sadomasochistic vertigo.

The operation proposed by R&Sie(n) could not be further from the idea of recycling, that expensive activity that extracts "clean" raw materials from scrap and refuse. Instead, this is a procedure of low energy consumption and a high level of conceptual regeneration, closer to ritual cannibalism or an extreme form of eroticized fetishism. It is a process of reversal that is grafted and turned upside down onto the similar metamorphosis indicated by Freud with the term *heimlich* (domestic, congenial, familiar), into its exact opposite.[12] We are looking at a chain of permutations in which defense of privacy, through various degrees of retreat from the gaze, gradually makes domestic peace secret, shameful (the latrine was indeed the *heimlich* room), mysterious, and disquieting. According to Freud, the appearance of what disturbs is an effect of the return of something repressed, something that was supposed to stay hidden and instead has unexpectedly returned to the surface.[13] So repression is the motor of anguish that transforms the desire of an irremediably past condition into an unavoidable threat.

The movement between the familiar and the disturbing, then, depends on the subject, his or her history, and the conditions that orient perception in one direction or the other. Where the object is concerned, the more domestic, normal, and familiar it is, the more sinister it can become: the house of Norman Bates, the protagonist of *Psycho*, is in fact a very peaceful wooden cottage that becomes frightening only due to the paranoia of the situation.

architects: R&Sie... year: 2003 location: Trinidad latitude: 10°04'N average temperature: 80.6°F (27.0°C)

recognition of the eroticism that results from nature conflicts and a built-in mosquito trap for a sadomasochistic negotiation with paranoia

client: Ema and Cesar Reyes cost: \$200,000 (€152,400) dimensions: 1,400 ft.^{2} (130 m^2)

life and death of mosquitoes in the double skin of the housetrap

< mosquito bottleneck >

General plan and views of the house

scenario

1)

Detection of the mosquito-borne West Nile virus on the island of Trinidad

2)

Mixing of this objective paranoia with the desire for safety

3)

Development of a Klein Bottle twist between the two contradictory elements: humans and insects

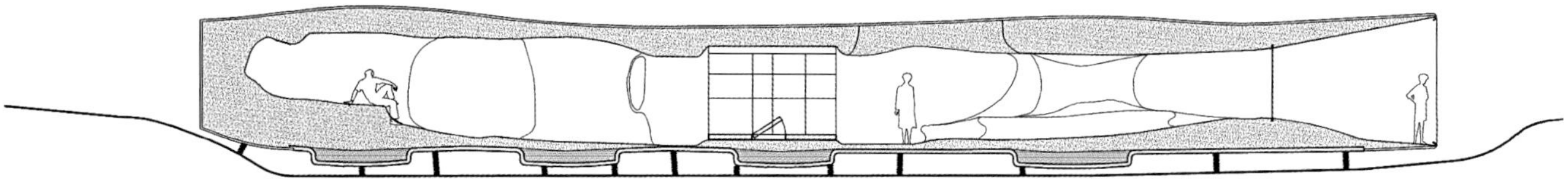

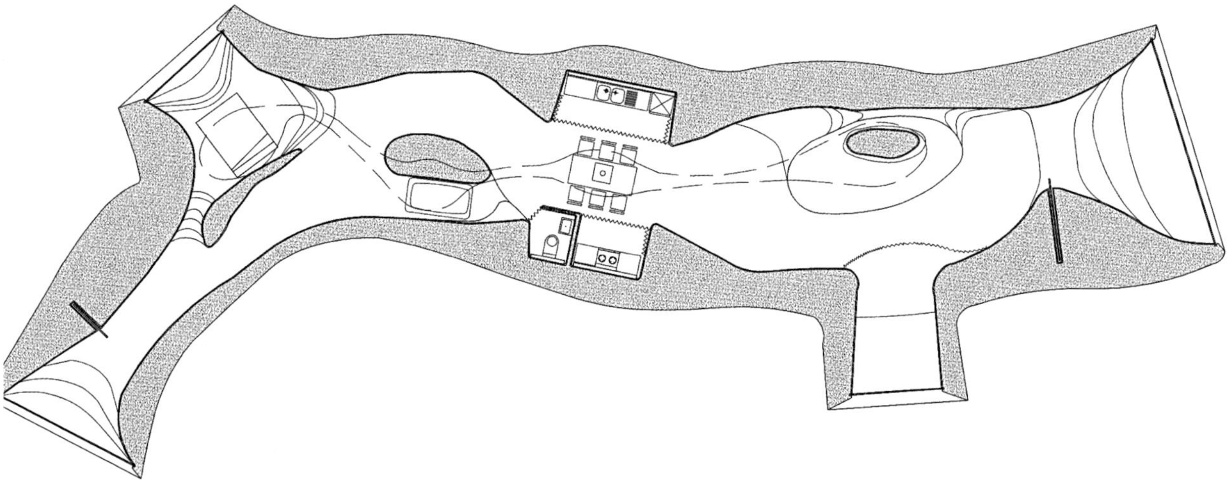

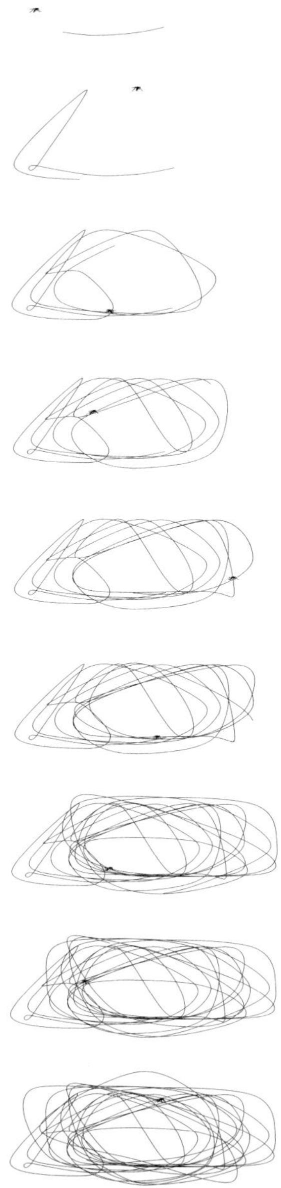

4)

Observation of the life and death of mosquitoes in the house trap

5)

Introduction of a fragile structure and materials (such as fabric netting) everywhere, in recognition of the island's geographical position (it is naturally protected against hurricanes)

Plan and sections:
The clients, an art-collecting couple, wanted a home for short vacations. The flight trajectory of the mosquito provides the form for the tubular spiral that supports the building and its double skin.

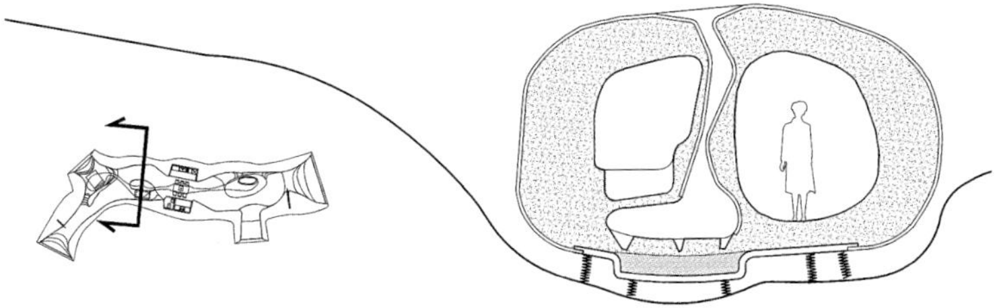

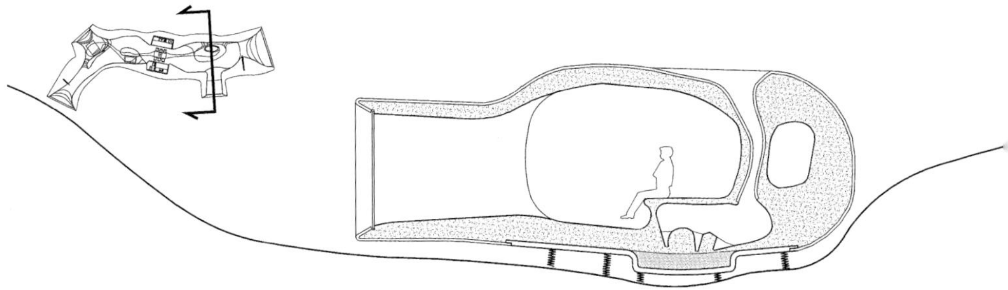

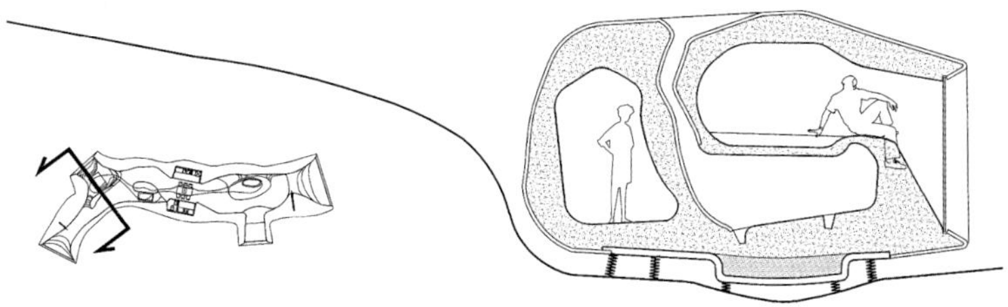

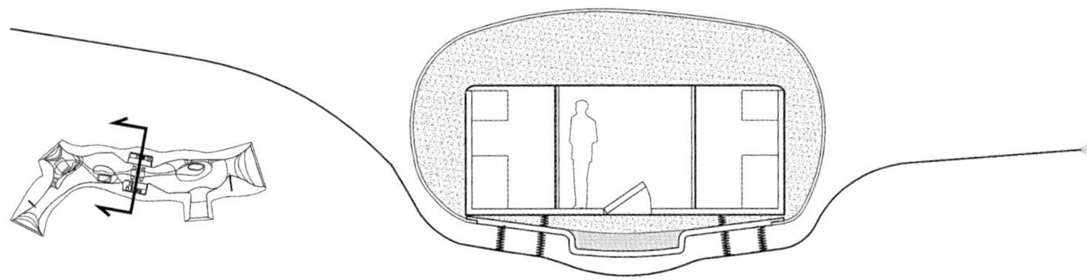

6)

Weaving together of all the surfaces of the house—floor, facade, and roof—with plastic wire and plastic shrink-wrap

7)

Resonance between the buzzing of the mosquitoes and the vibration of the structure

Detail of one of the entrances to the traps:
The mosquitoes are captured through the same principle used by local people who cut a plastic bottle and then insert the top part, turned upside-down, into the bottom part. The insects are attracted by the liquid but cannot find their way out of the trap.

Interior of the trap:
The choice to use polyethylene fabric, made with the same bottles used for the homemade traps, underlines the dual connection between the theme and its local interpretation.

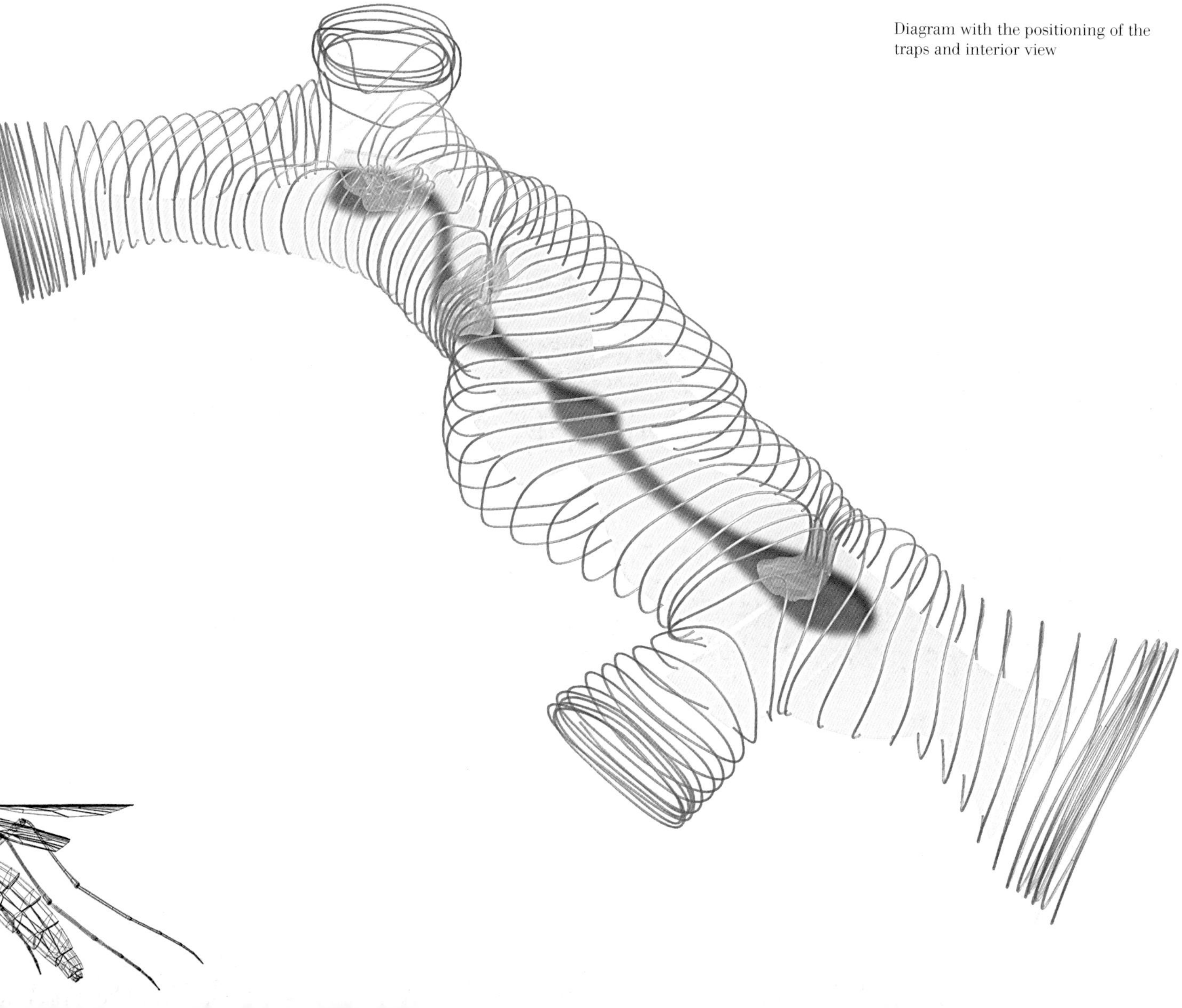

Diagram with the positioning of the traps and interior view

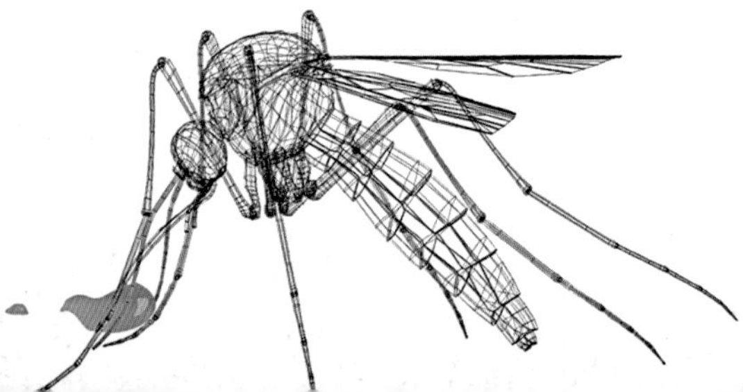

< multiple identity < utopia < fear and loathing < disorientation < furtive tactics < base materialism < danger

phantoms

The radical display of disturbing things, which Roche and partners put at the center of their approach, clearly works in a different direction with respect to a disquieting familiarity. Fears and desires here play a generative function, and the repulsive qualities of refuse become instruments of primary rooting in the specificity of contexts. Like the investigators of *CSI*, the architects rummage through trash and dissect cadavers to discover the most authentic and surprising realities, using the "evidence" thus gathered as tools of project activation. Participating in the competition for the new facility of the University Institute of Architecture of Venice (IUAV), they resolve their project **Aqua Alta**'s vertical composition with double walls of transparent plastic, which suck up the lagoon water along with the varied organisms and suspended matter that make it murky and smelly. In Bangkok, where they received the commission to design a new contemporary art museum (**B_Mu/ Dusty Relief**), they proposed covering the whole structure with a metal screen that is electrostatically charged to attract the dust from the relentless local smog. As is to be expected, this reckless return of the rejected, the depleted, takes on a spectral, phantomlike appearance in both projects. In the first case, this is obtained through the dissolving of an existing building's profile (which was to be replaced by the new university structure) in a liquid, greenish consistency of iridescent reflections. In the second, it is pursued by draping a stack of "white boxes," of exhibition spaces, with a wrapper similar to a classic sheet, whose dusty opacity affects the light filtered by the pollution, creating its own particular gray scale. Furthermore, the literal citation of breeding dust (the dust Duchamp allowed to settle and incorporated in *The Large Glass*), and its Venetian transformation into a similar accelerated cultivation of algae in the asphyxial soup of the lagoon, introduce a curious interaction with time and events. A continuous modification is triggered, an action of automatic contextualization that intersects with the lives of the buildings, above all after their construction.

architects: Roche, DSV & Sie   year: 1998   location: Venice, Italy   latitude: 45°26'N   average temperature: 55.2°F (12.9°C)

pollution and romanticism, extracted from the dirty lagoon through a sucking device

client: competition, University Institute of Architecture of Venice   cost: $6,000,000 (€4,572,000)   dimensions: 44,132 $ft.^2$ (4,100 m^2)

transparent capillary-effect pvc membranes

< aqua alta >

From the existing concrete to its ghostly replacement:
An ambiguous process of substitution, or crossing, occurs between what history has left us and its mutant recycling.
The vertical walls have a "wavy motion" when inserted in the existing volume.

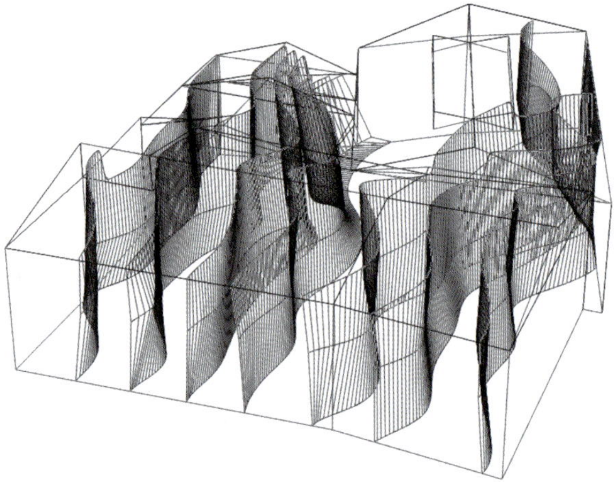

scenario

1)

Destruction of the existing cement warehouses (preserving a virtual "ghost print" of the old building)

2)

Digitized representation of water being sucked up into this imprint, with wave motions in each programming layer

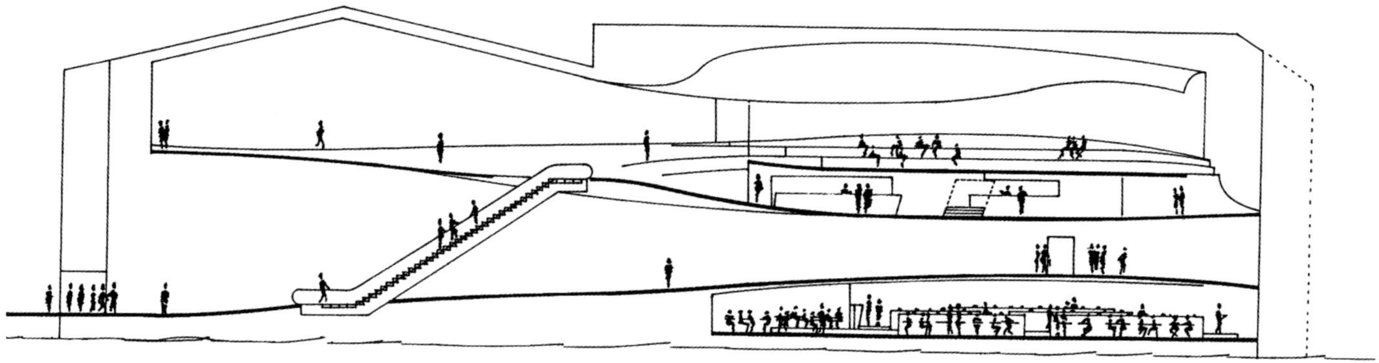

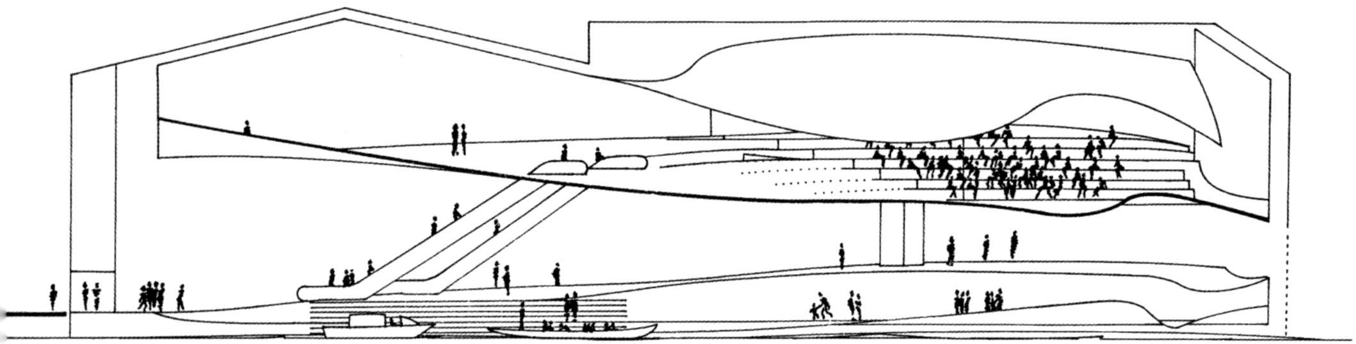

3)

Cutting up of the 3-D image to correspond to the imprint of the preexisting building

4)

Creation of internal and external PVC membranes

5)

Infusion of lagoon water vegetation in membranes through capillary action (into double transparent plastic walls for a high water exacerbation)

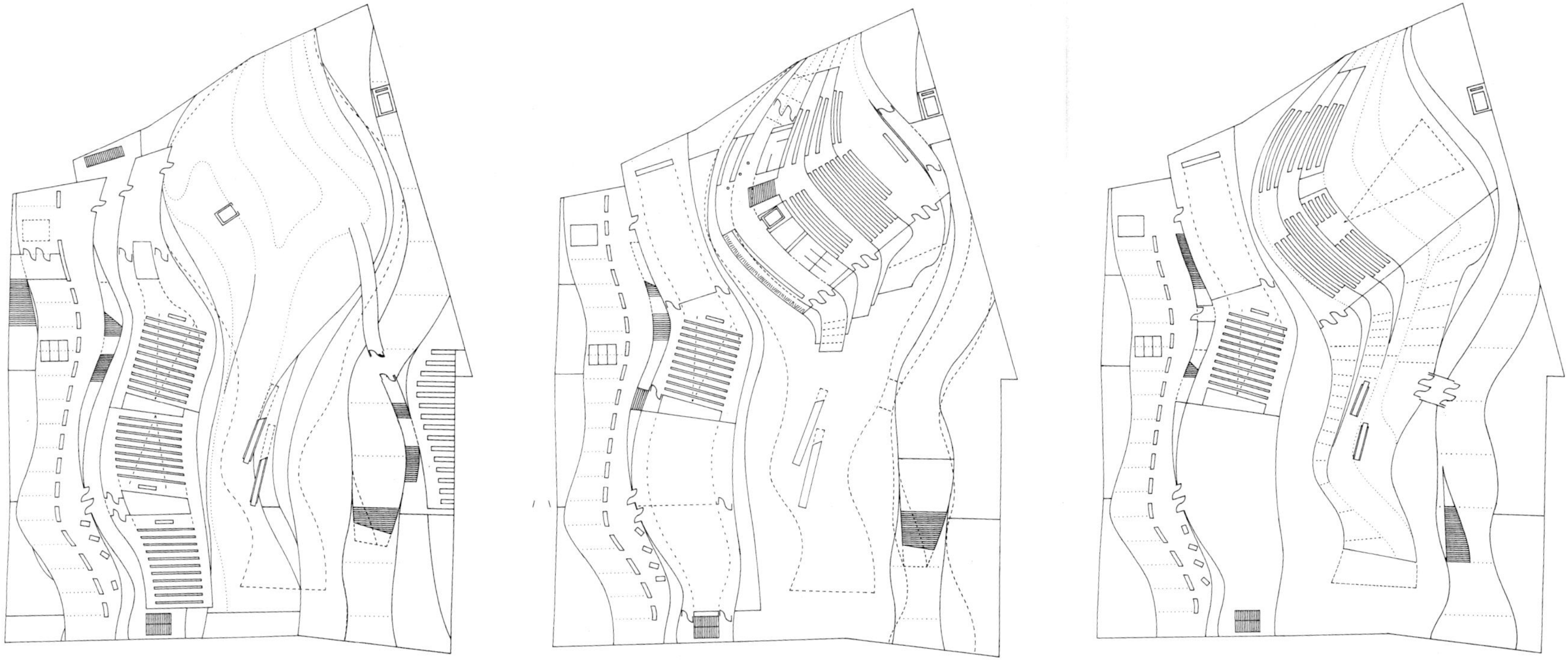

Sections and plans

architects: R&Sie... year: 2002 location: Bangkok, Thailand latitude: 13°44'N average temperature: 82.6°F (28.1°C)

pollution and identity are extracted from the dirty air through an electrostatic filtering device

client: Petch Osathanugrah, Bangkok Museum dimensions: 53,820 ft.^{2} (5,000 m^2)

electrostatic collection of suspended urban dust

< b_mu/dusty relief >

Dust cultivation, virtual model, and skyline of Bangkok:
Bangkok is both a gray and a luminous city, with a cloud of smog, a high percentage of carbon dioxide, and suspended dusts that filter light, whose spectrum seems to be limited to only the gray scale. A fog of particles makes the hypertrophic human activity of the city and its convulsive exchange of energies captivatingly visible.

scenario

1)

Calculation of random relief by a process of aleatory particles for a pure gray ectoplasm: This is under the lightning-gray sky of Bangkok.

2)

Collection of the city's dust on the surface of an aluminum latticework using an electrostatic system (100,000 volts / no intensity)

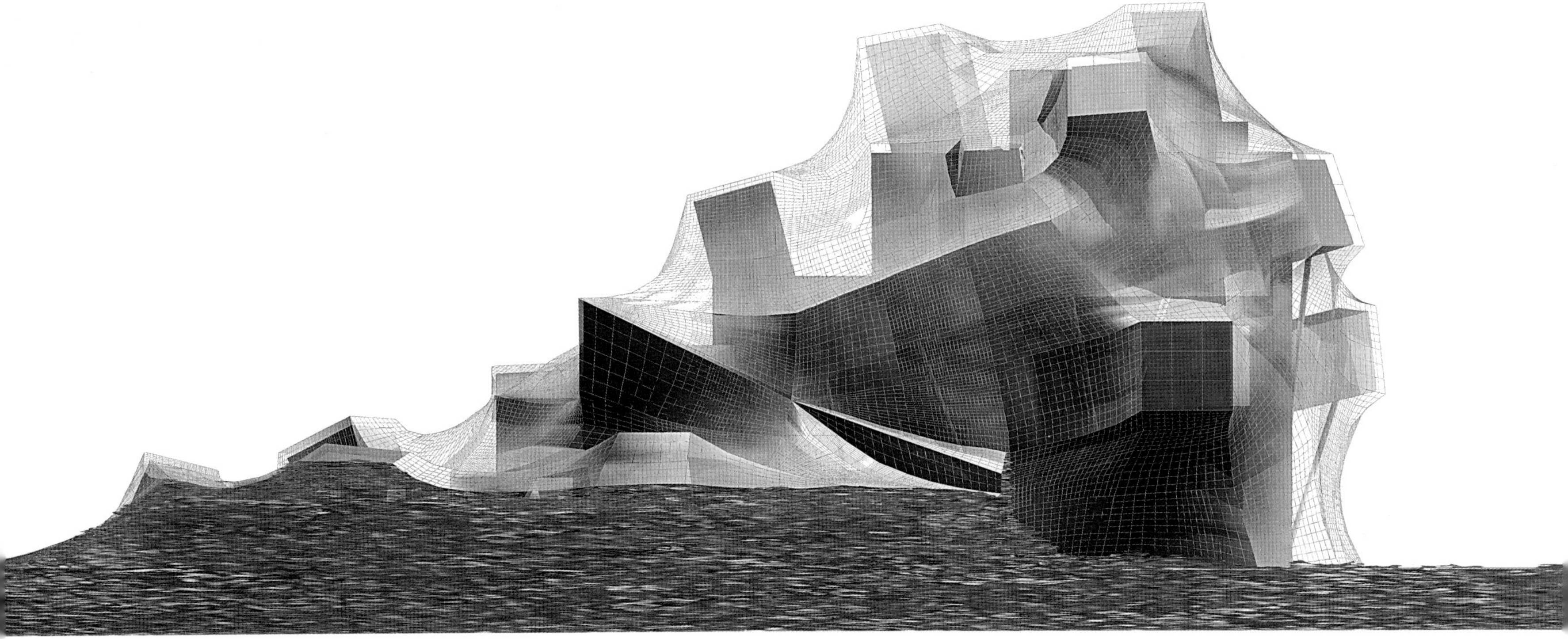

3)

Exacerbation of the "schizophrenic" climate, i.e. the contrast between the interior (white cube and labyrinth in Euclidean geometry) and the exterior (dust relief on topological geometry), used as an interface for an in/outdoor exhibition

Sections, interior view, and plans: The structure includes two distinct geometries: one Euclidean, which takes cultural merchandising and its worldwide circulation in aseptic, deterritorialized environments into account; the other topological, immersed in the toxic, mutant chaos of the metropolis.

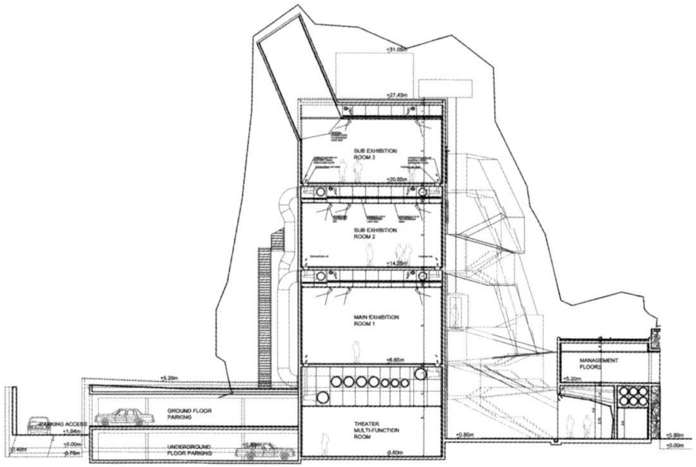

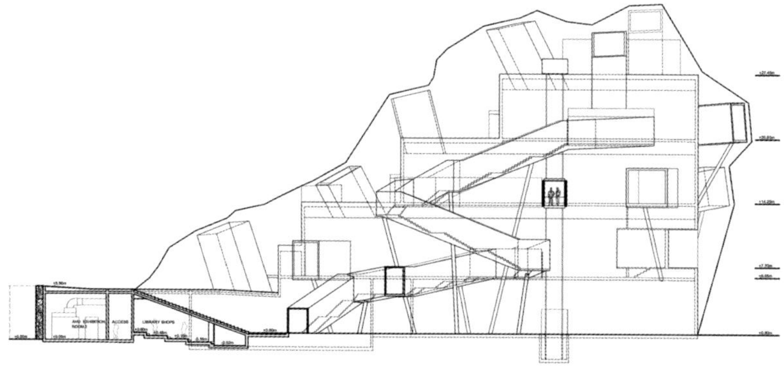

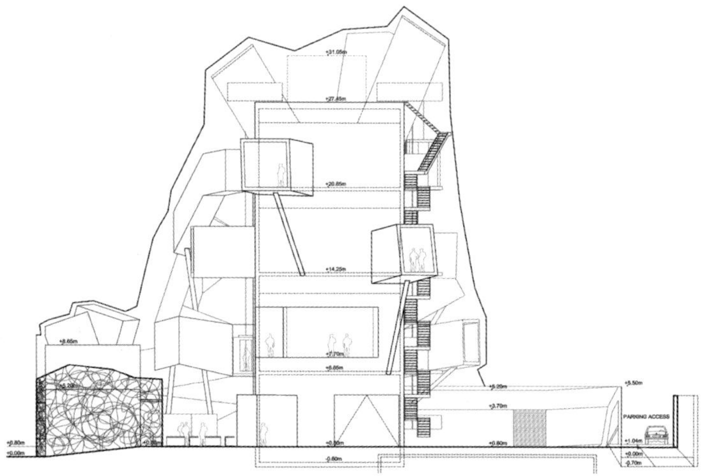

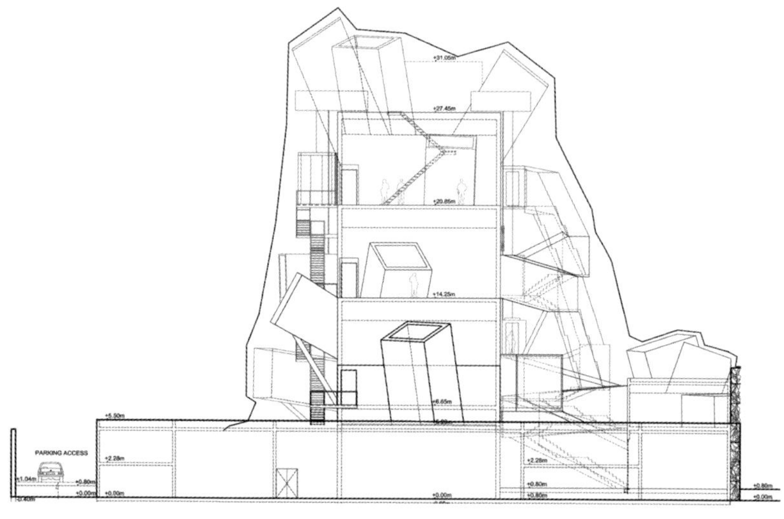

The B_Mu should function as a gate of local-global transit. The museum reflects the schizophrenia between the emergence of a phantom of the "cultivation of dust" of Duchamp as a strategy of hyper-localization, and the established "white cube" of the current museum typology.

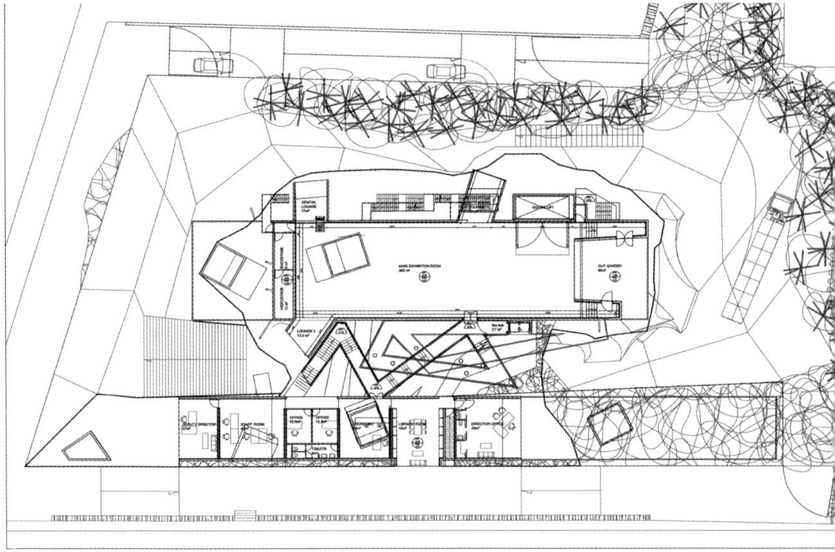

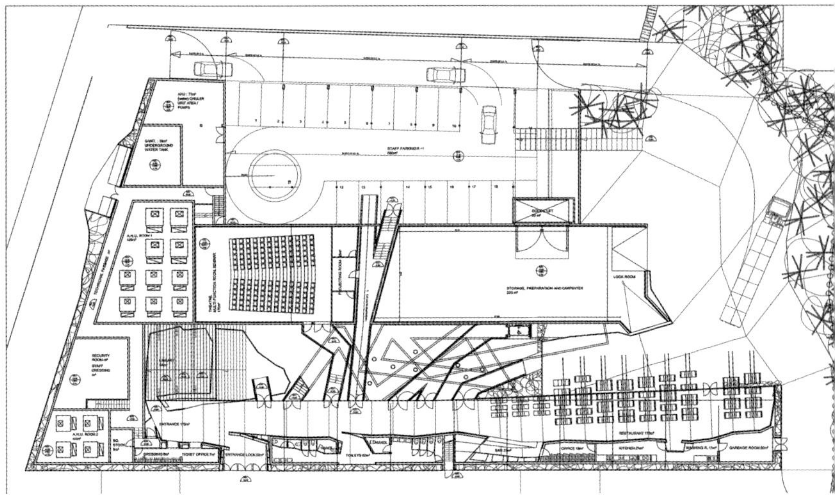

LIGHT
video
LIGHT
exhibition
access, access
video
access, access, access
exhibition
access, access, access
LIGHT
LIGHT
exhibition
video
movies
access, access, access, access
storage

Conceptual section and interiors:
The project feeds on the climate opposition between the overwhelming energy of the urban environment and the necessary procedures of climate control inside exhibition spaces.

< multiple identity < utopia < fear and loathing < disorientation < furtive tactics < base materialism < danger < phantoms

interstices

The temporal and interactive dimension explored in Venice and Bangkok, and further radicalized in **Olzweg** and **I've Heard About**, returns as one of the project operators of **Waterflux**, the glaciological and geological museum to be built at Evolène, in the Swiss Alps. This is a clone, more than an evolution, of the Scrambled Flat, an earlier proposal for the same site that called for the construction of an agricultural and residential complex. The idea was based on the radically remixed reinterpretation of the traditional stratification of stall, woodshed, beehives, dwelling, and hayloft. The connection with the specificity of the mountain biotope (in the sense of the age-old, economical-ecological symbiosis between men and animals) thus became the tool for an advantageous climate-control scheme for the building, exploiting the heat from the animals and the insulation from the hay. The various flows of nourishment of the livestock (grazing) and of fire (wood), together with the circulation of more-or-less warm and/or stale air, gave rise to a complex organic configuration inserted in a transparent enclosure whose form and size were based on those of a typical local rural house. Between this phantom parallelepiped covered by a double-pitched roof and the fluidity of the residential volumes, an interesting residual space formed that could be used to store hay.

What surfaces here is a sort of interstitial thinking that is found in many of the works of R&Sie(n). The double skin of the **Mosquito Bottleneck** is a very clear example, as are the snakelike tangles of the **Green Gorgon**, with vegetation that grows between edified ribbons; the spaces between the enclosure and the boxes of the museum, **B_Mu/Dusty Relief**, in Bangkok; the path that crosses and separates the density of vegetation and construction in **Spidernethewood**; or those that penetrate the glass of **Olzweg**; the suctioned lagoon water of **Aqua Alta**; the "cysts" of **(Un)Plug**. On the one hand, we see a reference to the complexity of natural organisms, the tangles of conduits, veins, and equipment (respiration, circulation, digestion) that separate and exchange the different bodily fluids. And also to the complex geometries of shells, cavities, curved and folded surfaces, determining situations of mutual articulation. On the other, a new strategy presents itself, operating on gaps, on the leftover zones generated by the schizoid intersection of programs that tend to conflict in terms of both space and meaning. A possible reference to the subtractive techniques applied by Rem Koolhaas comes to mind, starting with the project for the TGB in Paris, and then utilized in other proposals organized around programmatic separation. When used by R&Sie(n), these techniques infiltrate buildings with the potential of random events, of unexpected singular factors, anomalies capable of undermining any conventions regarding constructive and functional rationality.

This spasmodic pursuit of specificity is overlaid, at Evolène, by the paradoxical decision to keep the form and measurements of the building identical, in spite of the change of its purpose and the resulting variations in programmatic and symbolic content. This operation might, at first glance, seem like a return to the functional indifference of Aldo Rossi (whose typological-morphological approach, in the phantomlike citation of local houses, emerges in any case), or it might make us think, again, of Koolhaas, with his cynical transformation of a Dutch house that had remained on the drawing board into an enormous auditorium in Portugal. But the morphological permanence that resists a radical change in usage parameters can be more plausibly attributed to the antideterminist, cumulative thinking of Roche and partners, convinced that the acceleration of local conditions—which have remained the same—is more important than the solution of functional needs. Furthermore, the glaciological theme, with its connections to different physical states of water and their transitions, resonates with the "fluid" design dynamic that had given form and meaning to the earlier proposal. The building's interstitial spaces, defined on the outside by a porous texture, now accommodate atmospheric and seasonal variations, offering a series of different conditions for the meteoric waters and the sedimentation of signs made by changes in the conditions of humidity and temperature. The accumulation of snow in the winter in the same residual openings (intensified by a closed-circuit, artificial snowmaker fed by a small reservoir) makes the organic succession of the interior spaces a credible reflection of the cavities and crevasses that naturally take form in glaciers.

The specification of the program requirements is logically accompanied by an in-depth study of the constructive substance of the building. The complex internal volumetric arrangement, with its curved contours, is made by hollowing out large, prefabricated sections of solid wood using numerically controlled machinery.

Waterflux:
Traditional building in Evolène

The pieces are then glued together on-site. The resulting continuous surface responds—thanks to the qualities of the wood from a local species of tree—to the requirements of structure, waterproofing, insulation, and finishing of the museum spaces. This simplicity is achieved, surprisingly, by applying the most advanced technologies, through which the tendency to make building partitions out of more-and-more-specialized layers is radically inverted. This method's lack of detailing (no base or crown, no eaves, aprons, or gutters; no joints or transitions between materials, no difference between floor, walls, and ceiling) generates that ectoplasmic, monomateric consistency that seems to be the aspiration of many of the recent works of R&Sie(n). The fluid solidity of the wood reacts to the porous, semitransparent texture that contains the whole and reconstructs a sort of bodiless image of the traditional typology. Again, the texture resolves the public interface of the building, while offering intersecting relationships and connections, unlike other projects, between hard and soft, and between more-or-less-Euclidean geometries (see, for example, the **Shearing**, **Mosquito Bottleneck**, and **Spidernethewood** houses, and **B_Mu/ Dusty Relief**). What remains, in any case, is the idea of materializing something directly from the context, like the sand monster of Spiderman or the golem of Jewish tradition, exploiting the continuity with basic conditions (see, for example, the hydroponic cultivation on the surface of the **Green Gorgon**, utilizing the same plants found in the lakeside surroundings, or the "suctioned" lagoon of **Aqua Alta**), with existing construction (the "foam" of the **Mi(Pi) Bar**, extruded from one of the sides of the building by Pei at MIT and made with the same ceramic material), with local industry and the virtuous behaviors of residents (the recycled glass in Orléans that also covers the mechanical arm, making it identical to the greenish blob), or accommodating atmospheric events, the pollution of Bangkok or the "immaculate" snow of Evolène.

Waterflux:
Crevasse in a glacier

architects: R&Sie year: 2002 (winner of the competition) / 2007 (restarting process) latitude: 46°06′N average temperature: 43.2°F (6.2°C)

nostalgic-industrial transformation of wood in an icy climate for a simulacrum of tradition

location: Evolène, Switzerland client: Fondation La Maison des Alpes cost: $5,000,000 (€3,809,500) dimensions: 9,386 ft.^{2} (872 m^2)

transformable museum of icy snow and lamellar wood

< waterflux >

Virtual model and scale prototypes of wooden shells:
The project is made completely of lamellar wood of a local species (Mélèze) in bundles twenty inches (fifty centimeters) wide, milled by a numerically controlled machine. The resulting parts are used for the structure, the insulation, the waterproofing, and the interior/exterior finishes.

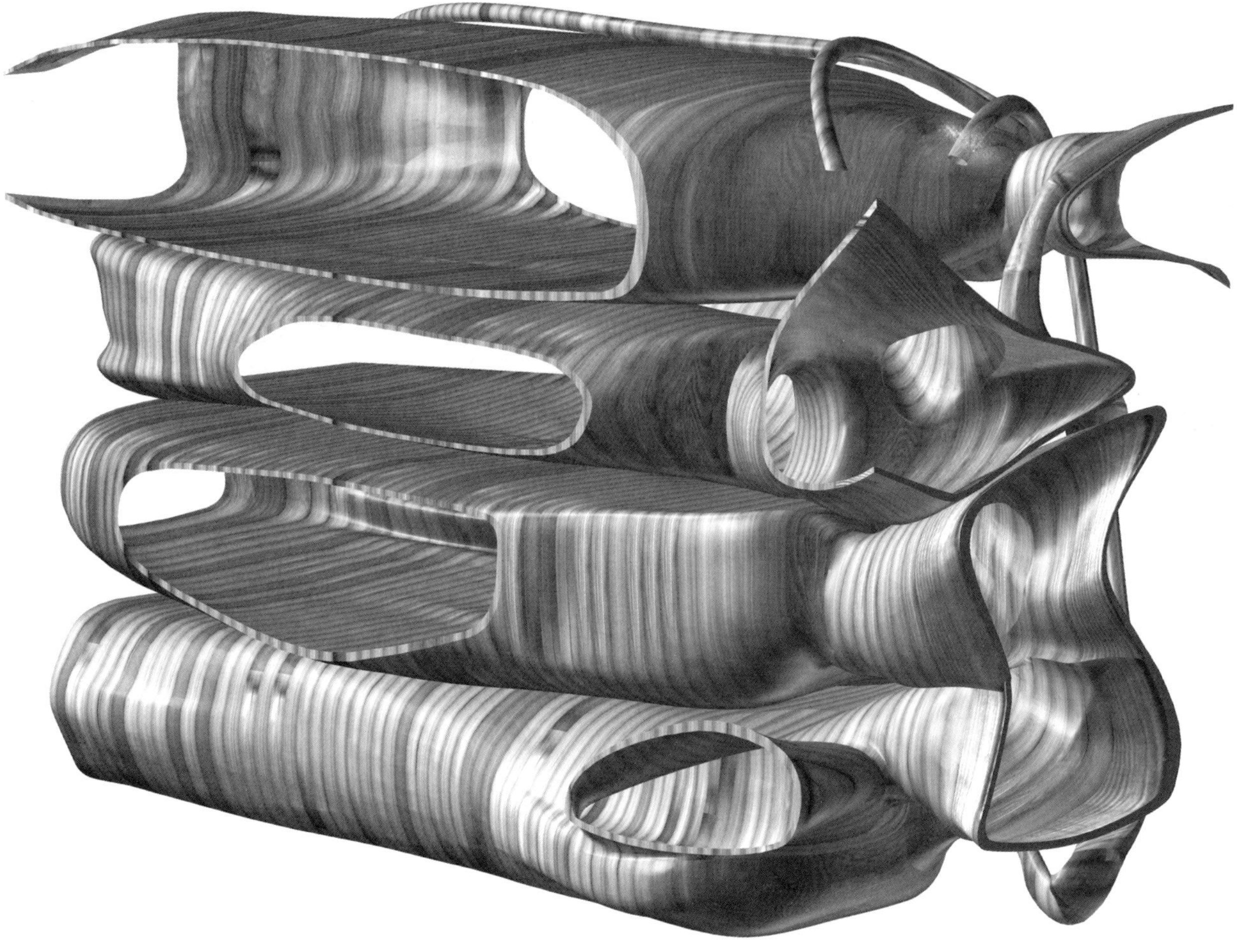

scenario

1)

Digitization of the envelope of a traditional habitat

2)

Scooping out of hollows within this volume, as if it were an ice cavity (but of solid wood, with a five-axis drill machine)

3)

Water states vary according to the seasons: the ice flows and freezes; the ice facades freeze and melt, forming a pond in front of the building.

Longitudinal sections:
The glacier museum makes the geological and meteorological environment of the high mountains visible, offering refrigerated spaces for art installations and scientific demonstrations.

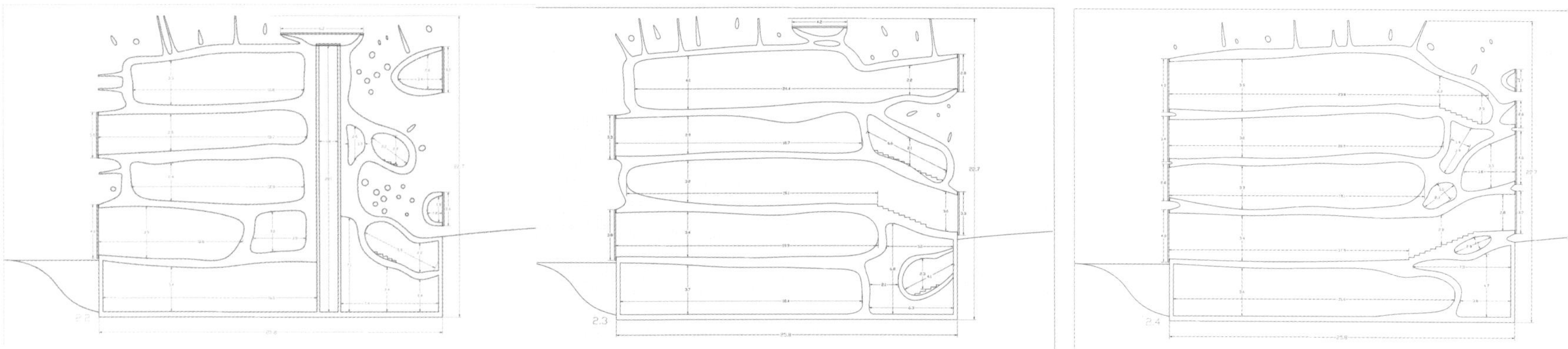

4)

Exacerbation of the winter climate with 17,650 ft.^{3} (500 m^3) of artificial snow

5)

Construction by CNC machine processing out of 1,000 trees' worth of solid wood (70,600 ft.^{3} / 2,000 m^3) and reassembly of the 180 manufactured pieces on-site

6)

Rejuvenation of the local economy

Cutaway of virtual model:
The building is wrapped with a grille (opacity 15 to 20 percent) that reproduces the profile of traditional houses, an enclosure that makes it possible to hold the snow inside a phantom typomorphological imprint.

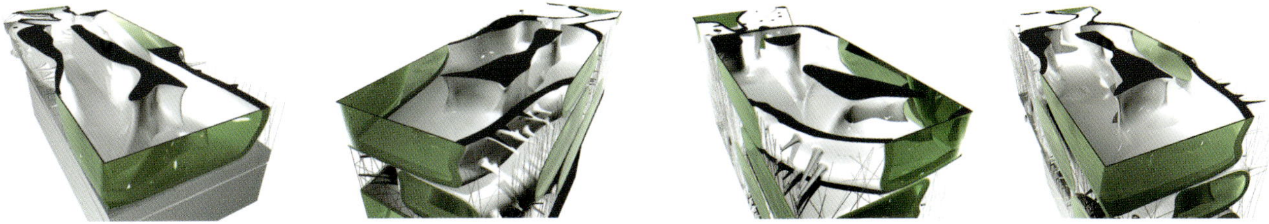

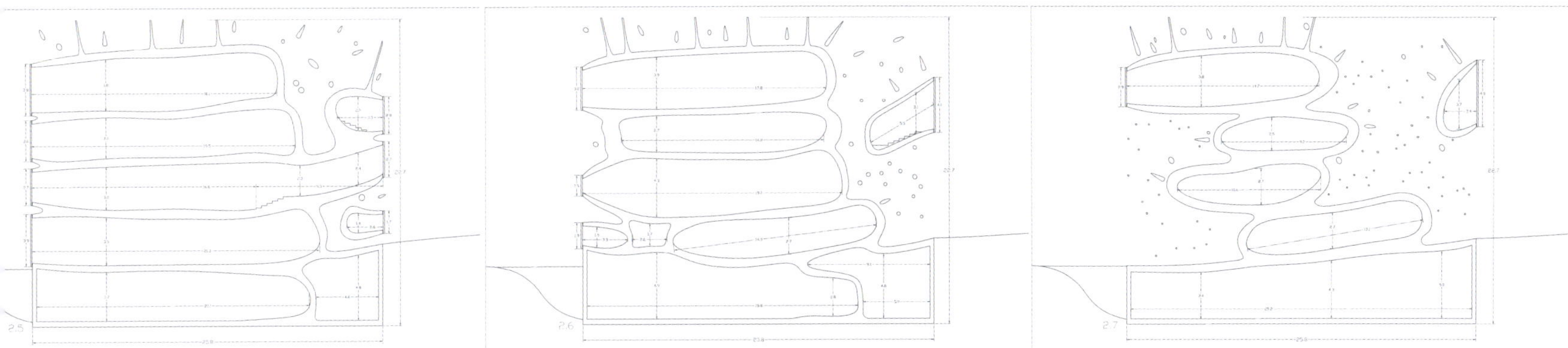

Virtual model:
The transformable building reacts to the rhythm of the seasons. In the winter it appears like a solid cutout of ice and snow, with cavities similar to those found in glaciers. In the summer it resembles the piles of stones used in these areas to mark borders. A small pool collects rainwater and supplies it to the artificial snowmaking system. The phases of transformation of the water thus become an integral part of the building.

< multiple identity < utopia < fear and loathing < disorientation < furtive tactics < base materialism < danger < phantoms < interstices

machines and narratives

To bring latent, inert, or entropic materials to life, Roche and partners make use of a tool that is simultaneously simple and sophisticated: writing. Of course, we are not talking about the magic spells wise rabbis traced on the foreheads of their slaves of clay (or on little strips of paper to have them swallow), nor is this a matter of obtaining a superhuman but stolid, submissive force. The infidel golems of R&Sie(n) are in fact possessed by the most advanced cybernetic applications and activated, rather than controlled, by software in which lines of code permit complex negotiations with local conditions. The outcome is an architecture that increasingly intersects with robotics, both in the conceptual terms of a sensitive device capable of perceiving its environment and extracting/infiltrating strategies of mutation, and in the more concrete terms of a growing mechanical presence crossed, like a hybrid, with architecture itself.

In **Waterflux**, for example, the computerized simulation of flows of heat and nourishment provides the initial stimulus for a morphological solution, and is the prelude to the intervention of an innovative CAD/CAM machine capable of making large wooden parts of complex geometries (up to a maximum of 131 x

16 x 1.6 ft., or 40 x 5 x 0.5 m). A similar construct of clashing flows was the basis for design exploration in **Loophole**, while other stochastic and interactive scripts underlie the variable morphologies of **Olzweg** and **I've Heard About**. This has nothing in common with the repetitive production of the Taylorist assembly line (Henry Ford sold his Model T in "any color, so long as it's black"), nothing to do with the "friendlier," flexible system today that allows customers to choose their own squalid car from thousands of different combinations of paint, upholstery, motors, and so on. Instead, the aim of the machines invented by Roche and partners is to be bio-techno-disturbing tools of hybrid constitution, possessed by a ferocious, erotic force—proliferation vectors of subjectivity and desire, rather than instruments of serial reproduction of a single prototype.

Another interesting link to the intuitions of Duchamp emerges, in this case with the powerful sexual symbolism unveiled by the prosthetic objects of industry. The improbable encounter between molds, pistons, grinders, and other strange readymades that becomes the dance of the bride and her bachelors in *The Large Glass* posits the ascendancy of the pleasure principle over that of production, the eroticism that is liberated by separating the act from its productive ends. This perverse use of the machine translates into a design practice that transforms artistic nonsense into a paradoxical operative tool: the rigidity and arbitrary quality of the rules lay the groundwork necessary for the automatic generation of alternatives. **Hybrid Muscle**, for example, is a project suspended between conceptual installation and supply of services, both a desiring machine and an energy system. Done in collaboration with Philippe Parreno at the Land Foundation—the utopian artistic-agricultural project launched by Rirkrit Tiravanija in northern Thailand—this strange device provides shade, shelter from the rain, nocturnal lighting (in the tropics the sun sets at six all year round), and electricity for visitors and students from the university of Chiang Mai, who can recharge their mobile phones and laptop computers.[14] The local resource that gives sense to the operation is animal traction: an albino water buffalo (it was supposed to be an elephant) raises a weight whose energy potential is then recouped by means of a generator.

Another albino animal, an electromechanical penguin, is the protagonist for an installation R&Sie(n) and Pierre Huyghe proposed in 2006 for the MAM in Paris and the Tate Modern in London. The stimulus for this work is the anomalous surfacing of a dark **Terra Incognita** recently exposed by the melting of ice caused by climate change. This dark surface reveals albino penguins, animals normally cast out of the herd and condemned to certain death in the freezing Antarctic wind. The topography of the place is reconstructed, like a gigantic pop-up, by making a series of cuts on a sheet of aluminum honeycomb that attains a third dimension thanks to a system of cables, held in place by counterweights filled with water. The gradual, random evaporation of the liquid during the period of the exhibition simulates a process of accelerated erosion, leading to inevitable flatness.

Here, as in all the works of R&Sie(n), we see two forms of writing that feed on each other, one of a technical nature (the "objective" interpretation of the place, its specific ecology, the script behind the three-dimensional modeling), and one that narrates. They give meaning to the automatized operations, literally producing a series of stories. Science and fiction form a hybrid, a sci-fi narrative capable of weaving an interactive relationship with reality, penetrating it, modifying it.[15] The strategy acknowledges the gradual blurring of disciplinary boundaries and, at the same time, the parallel and cumulative advance in greater depth of different areas of expertise, which tends to block actual collective knowledge. To tell stories, to reveal parallel universes, to explore the most improbable paradoxes, to infiltrate the media, to manipulate perception of reality through narrative—all become necessary tools for modification of reality itself, to create the conditions in which the heretical architecture of R&Sie(n) can find favorable terrain for growth.

One evident clue of this sci-fi attitude is undoubtedly the "scenario" used as a pithy introduction to each project of R&Sie(n). The "what if and then" device typical of this narrative approach joins interpretation and project, places and processes of transformation, together in a system open to mutation, to the proliferation of stories and their continuous falsification. A liquid, uncertain instrument capable of coming to terms with the widest range of manifestations of the contemporary world, whose most specific terrain is that of urban operations where the complexity of the parameters involved, their quantitative weight and temporal dynamics, heighten the degree of indeterminacy.

architects: R&Sie(n) / Pierre Huyghe   year: 2006   latitude: 72°28'S   average temperature: 33.8°F (1.0°C)

fading topography and an animatronic penguin from antarctic stories

location: Isla Ociosidad, Antarctica; Musée d'Art Moderne de la Ville de Paris; Tate Modern, London   dimensions: 2,153 ft.^{2} (200 m^2)

global warming and the apparition of the island of the albino penguin

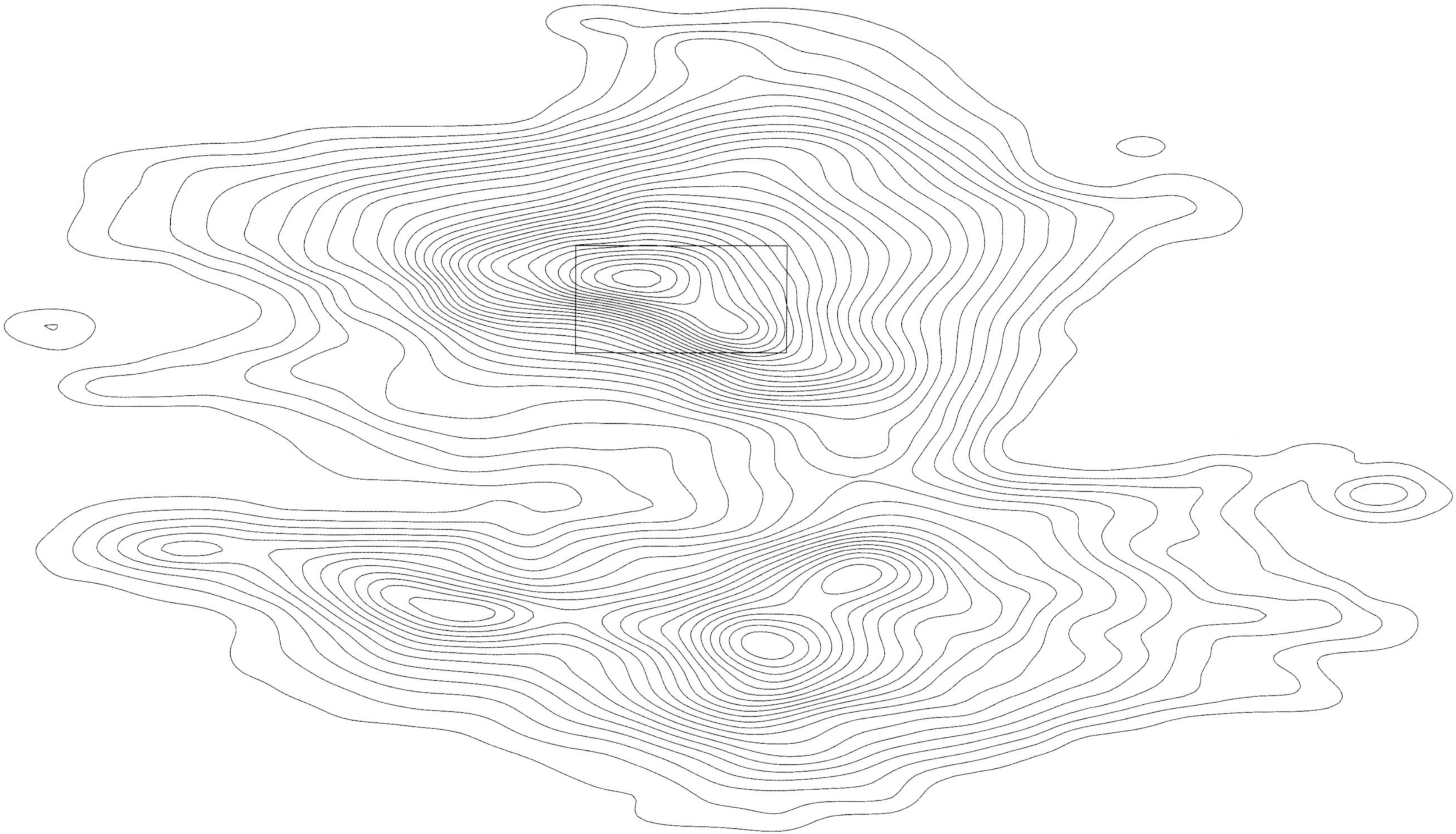

An odyssey in three steps

scenario 1)

Survey of the real situation of the site in Antarctica

Expedition to the Antarctic:
Terra Incognita is an island that has recently emerged as a result of global warming and the melting of polar ice. The albino penguins, exiled from their rookeries because they are different, become immediately visible against the dark rocks that have just appeared.

2)

Scanning of the territories that emerge once the snow has melted

3)

Redevelopment of the instability process in two different places, namely the Tate and the MAM

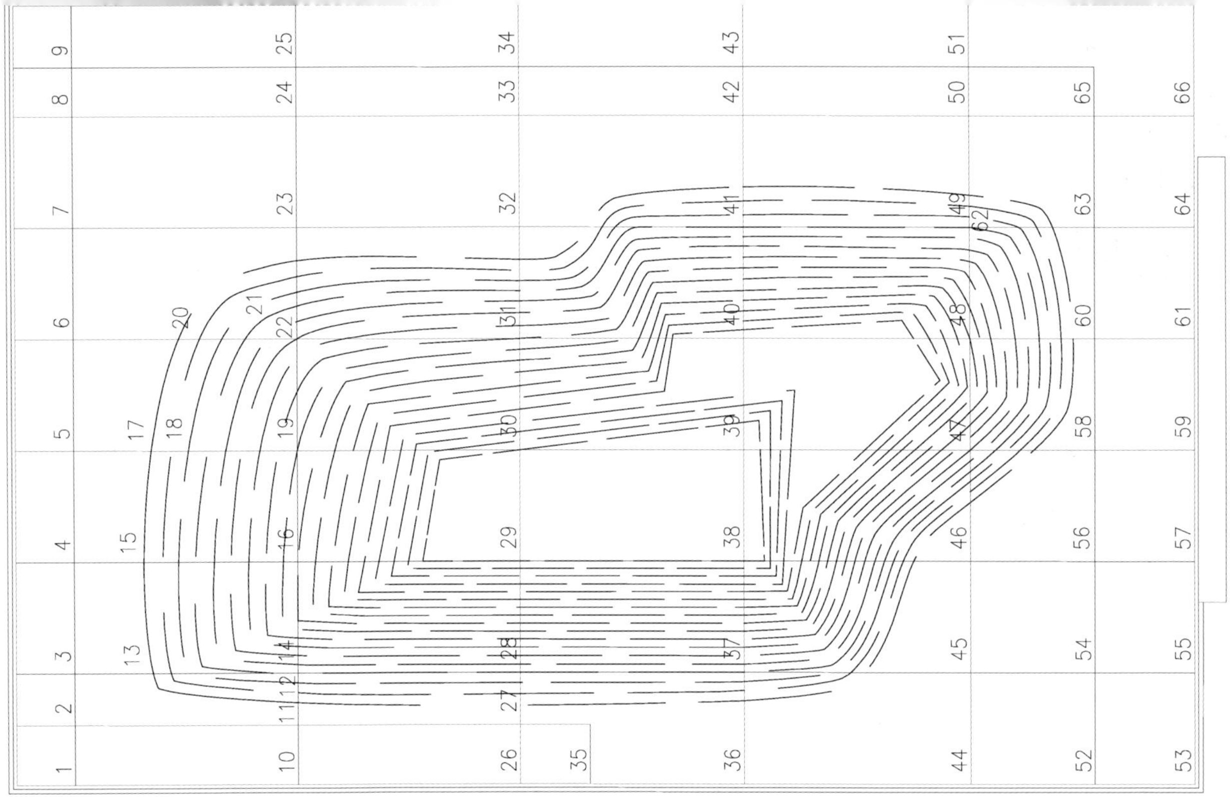

Working drawing of the aluminum honeycomb surface, with indication of the cut segments and individual sheets: The installation is composed of a reproduction of the topography that has emerged from the ice, through a surface made of cut and raised honeycomb aluminum. A numerically controlled machine that uses high-pressure water jets makes the cuts.

Assembly of the aluminum honeycomb surface:
The topography takes form thanks to a series of cables that raise the interconnected cut sheets of aluminum. The counterweights are filled with water, a reference to the melting of the ice. As the water evaporates the artificial island redefines its flatness.

View of the installation, formed by suspended segments of honeycomb aluminum and a robot model of the albino penguin

architects: R&Sie... / Philippe Parreno year: 2003 location: Chiang Mai, Thailand latitude: 18°46′N average temperature: 78.6°F (25.9°C)

animal paganism and a power station for a contemporary ceremony place

client: The Land, Rirkrit Tiravanija cost: $65,000 (€49,530) dimensions: 1,400 ft.^{2} (130 m^2)

animal traction generator

< hybrid muscle >

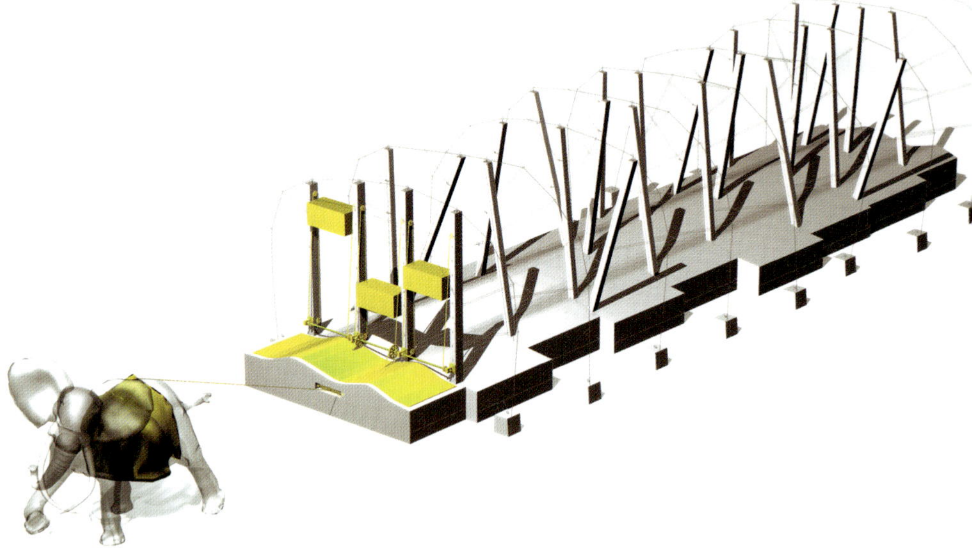

scenario

1)

Construction of an animal "engine" driven by the muscular power of an elephant: Mechanical energy is stored through the lifting of a twenty-ton steel counterweight. The mechanical energy transforms into electrical energy. This can power ten light bulbs, laptops, cell phones.

2)

Natural ventilation through the pneumatic-aided fluttering of the facade, made of sheets of elastomer that work in the same way as temporary shelters made of teak leaves

Diagram of the machine powered by an elephant, and the completed structure

post scriptum:

An albino buffalo replaced the elephant (the ground of the rice field is too fragile for a pachyderm).
The counterweight was reduced to two tons in only one location (for security reasons).
The pneumatic movement of the plastic leaves has been abandoned (for a decrease in cost).

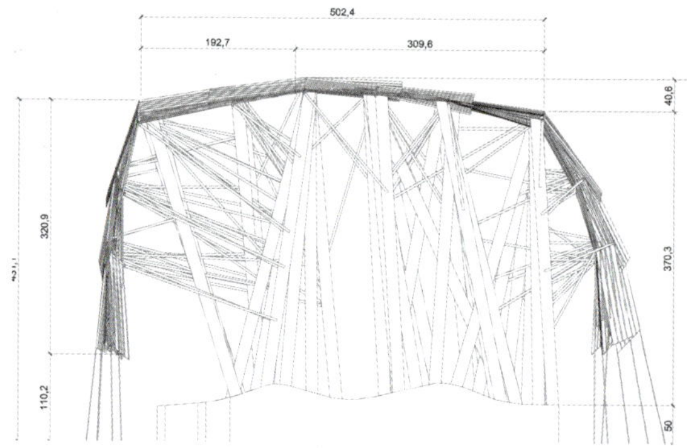

Views and cross section:
The albino buffalo replaces the elephant indicated in the preliminary project.
The two-ton counterweight is raised twice a day.

Plans and section of the pavilion and construction diagrams of the electrical generator

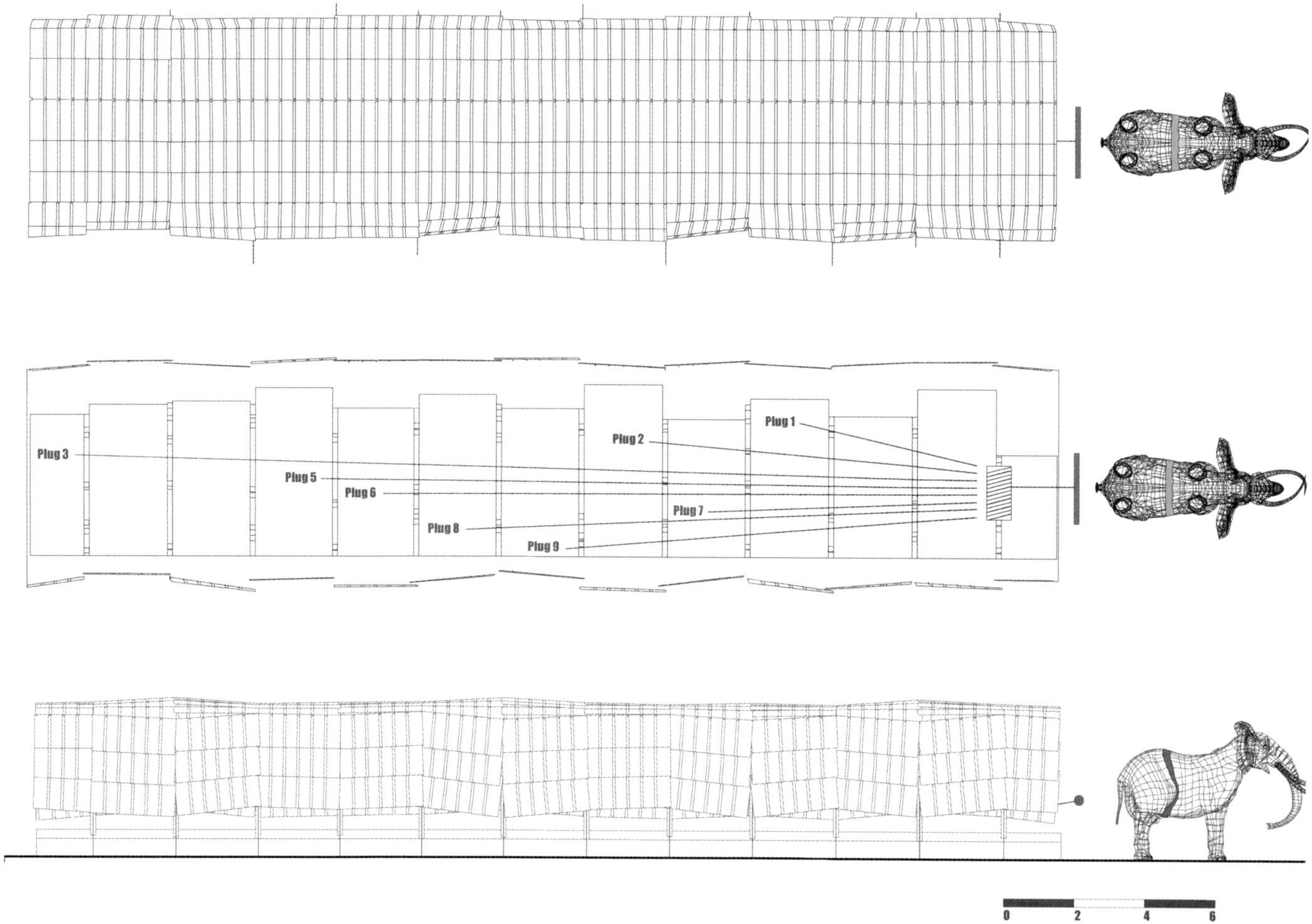

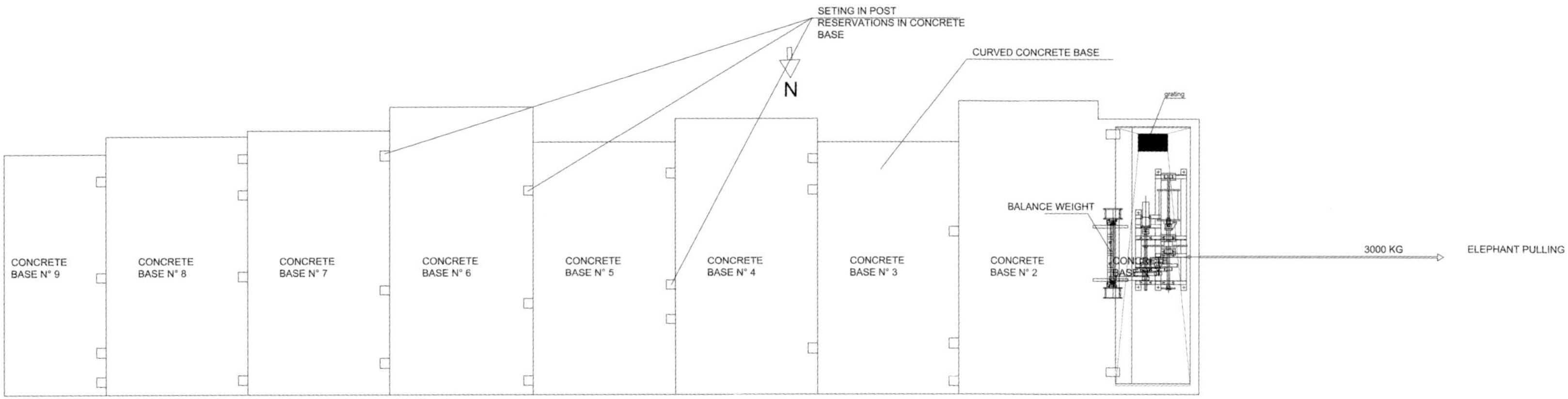

PLAN OF FORCES OF BALANCE WEIGHT

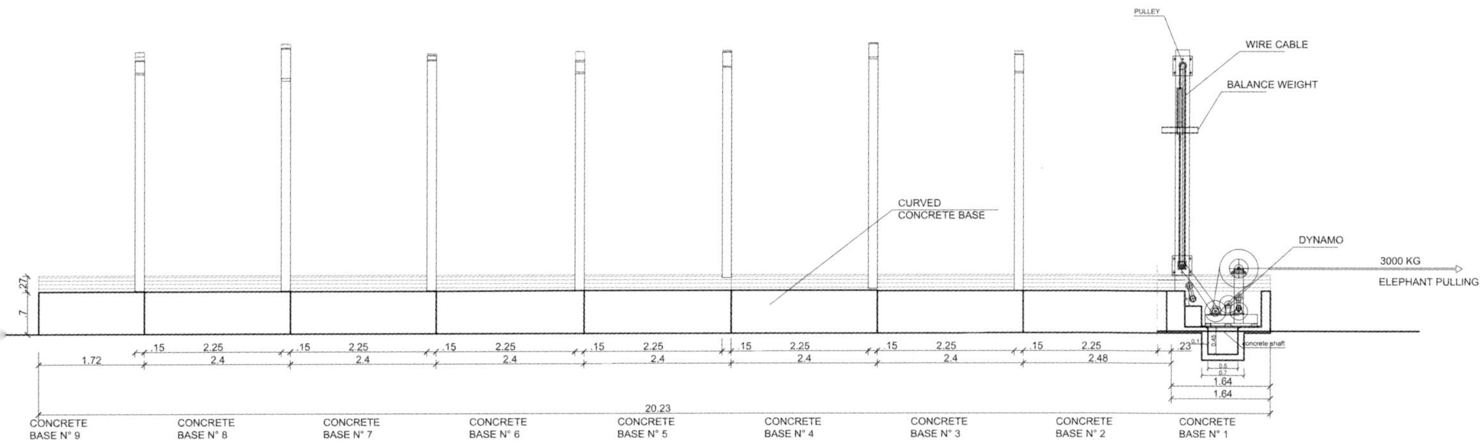

NORTH FACES ELEVATION

< multiple identity < utopia < fear and loathing < disorientation < furtive tactics < base materialism < danger < phantoms < interstices < machines and narratives

toxic ecologies

Swarm Town, the project submitted for a competition for a new city of 494 acres (200 hectares) in South Korea, shifts the automatic, robotic, and interactive methods of **I've Heard About** from the terrain of a sociospatial utopia to an approach to a specific situation, namely a place characterized by an overwhelming physical quality. Nature becomes the protagonist of the scenario, the active subject of the project and, at the same time, the object to be protected. Faced with the by-now-inevitable theme of environmental sustainability, the corrupt, paranoid vision of R&Sie(n) chooses to operate between the lines, concentrating on a radical intensification of the premises. From landscape romanticism and conservative environmentalism—ideologies that ensure the "political" feasibility of any large-scale settlement operation—the group extracts a series of strictly quantitative constraints that are managed through a process of parametric design.

Once again, the taking of local situations to extremes becomes the vector of transformation, generated by three different scripts that operate in succession. The first addresses the dynamic of flux inside a grid deformed by the existing topography. The program takes different modes of movement and interaction with individual behaviors into account (simulated through a series of virtual "particles"). The next algorithm inserts quantities of construction compatible with their purposes, the mobility system, and the topography. The variations of "pressure" recorded by the software generate a series of local hypertrophies in which to concentrate densities, with a system of cross-checks on the functional compatibilities and the resulting skyline. The third script works on the interior-exterior interfaces, and, thanks to spider robots that range over the enclosures, triggers a process of local characterization that works in terms of "difference and repetition," in pursuit of an open, variable standardization. These electromechanical agents intervene in the life of the city right from the moment of its construction, with the assembly of the external surfaces, the organization of hydroponic cultivation, and other botanical coverings. In the following phases, they also ensure operations of management, cleaning, and maintenance, and the necessary transformations.

The open, interactive character of the program permits rapid feedback between the requests of the client and the design process. The latter, mostly entrusted to autonomous, sensitive, generative agents, is similar in conceptual terms and physical results to the geological events and the successive forms of erosion that have formed the Korean site. Iteration of the three scripts also focuses on possible solutions with the parameters of environmental compatibility, linked to reduction of emissions, energy accounting, growth of vegetation, and mobility, calculating individual and collective ecological footprints at each passage.

Among the projects of R&Sie(n), **Swarm Town** has perhaps the greatest focus on sustainable development, taken as a critical node of the postcapitalist "pensée unique." We can clearly sense the appeal of the widespread fear of environmental disaster for François Roche's paranoid thinking, along with the need to sabotage the mechanisms of its commercial exploitation. Sustainability, though far from having been achieved in real terms, is in fact one of today's most effective propaganda devices to make operations that would otherwise be rejected for many reasons (including real environmental concerns) seem more acceptable. The result is an image bank in which fear and faith coexist, nourished by the catastrophic visions of Paul Virilio but without his millenarist tones, where the "day after" becomes a possibility to explore, a condition that is already present, here and now. The hermaphrodite polar bears

of the Svalbards or the Jurassic tree of *Charisma*, as weak as it is threatening (both always on hand at the latest lectures of François), confront us with the complexity of the negotiations, the "natural" conflicts and adaptations, while helping to identify the many niches in which to cultivate new design research.[16,17]

In **Thegardensofearthlydelights**, the very recent "poisonous" garden created for the Dalmatian island of Lopud, the mobile threshold between the toxic and the therapeutic is resolved through the paranoid automatism of the positioning of various species, ranked by their level of danger. Another example, but on a much larger scale, is **(Un)Plug**, the proposal for an "ecological" skyscraper commissioned by EDF (Électricité de France), which explores the paradox of constructing the headquarters of the French electrical utility to be a self-sufficient building with the potential to avoid the services the company supplies: a tower clad in fur (composed of thousands of transparent tubes in a vacuum, for heating) and marked by protuberances (covered with photovoltaic panels), a mixture of Meret Oppenheim and Matthew Barney.

What also emerges through these scenarios, aimed at undermining the clichés of environmentalist propaganda, is the tendency of Roche and partners' design devices to let themselves be corrupted by contexts, while revealing their contradictions. They have the capacity to negotiate with the most complicated, problematic realities, to fully accept situations, and to derive innovative mutation strategies, in which the elusive, changing identity of the group and its curious, interrogative approach play a fundamental role.

In a design panorama of increasingly rigid conformity to reassuring, standardized solutions, the extreme, perverse, aberrant gaze of R&Sie(n) triggers a highly liberating action of disturbance, pointing to anomalies, and, above all, the construction of conditions that make them possible, as the necessary objective of architectural research.

Swarm town:
Flow scripts, automatic interactions between flows and topography

architects: R&Sie(n) / Benoît Durandin year: 2006 location: Chungcheongnam-Do, South Korea latitude: 36°29'N

swarm intelligence, random agents, and stochastic scripts for an unpredictable, interactive urbanism

client: competition, Multifunctional Administrative City Construction Agency, government of Korea dimensions: 21,527,800 ft.^{2} (2,000,000 m^2)

sustainable development, critical node of the capitalist future

< swarm town >

A cycling path

scenario

1)

Acknowledgment of the site's features, mainly influenced by its natural topography, as a form of pressure: the revenge of nature on planning

2)

Assumption of sustainability as the main issue: an urban settlement as a contradictory, collateral effect of the green area

3)

Introduction of a generative and interactive process, fully automatic and objective, not ideological, taking into account geographic, programmatic, and environmental parameters

Swarm Town's urban morphology comes from the interaction of three scripts: the Flow script, a generator of urban grids; the ZXY scalar script, involving processes of densification; and the Nerves script, connected with the management of internal-external interfaces.

These three processes introduce a system of interactive variations connected with the following parameters: survey (geography and topomorphology of the place); form (height and volume of the buildings); and programs (as the iterative, inductive, and deductive reinterpretation of functional requirements in relation to spatial results).

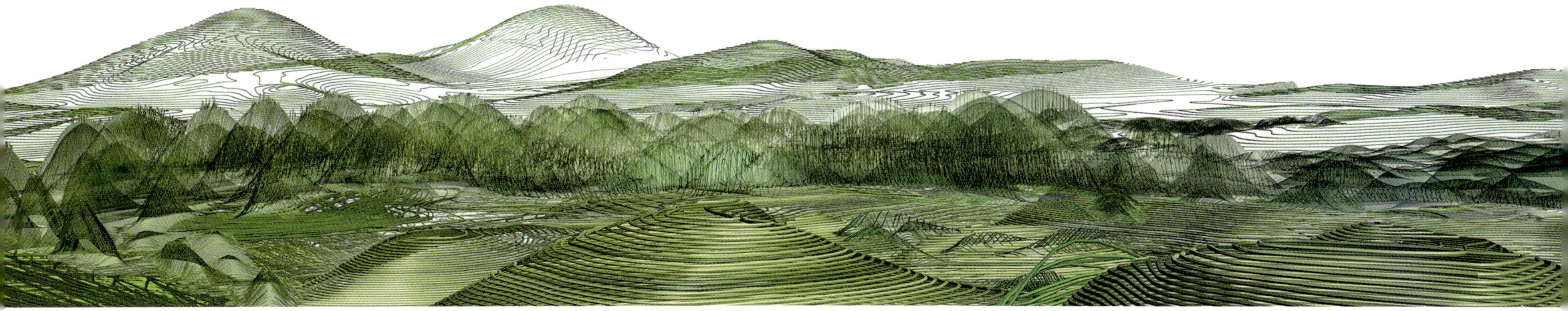

4)

Algorithmic and dynamic transformation from embedded parameters to shape the urban physicality

5)

Real-time check of the feedback effects on ecological footprints and the organization of the town

ZXY scalar script:
A new inhabited topography emerges from the interaction between flows and the characteristics of the terrain. The urban forms are determined by automatic procedures directly connected to local situations. The genius loci is no longer a "green" alibi but a vector of urban design transformation.

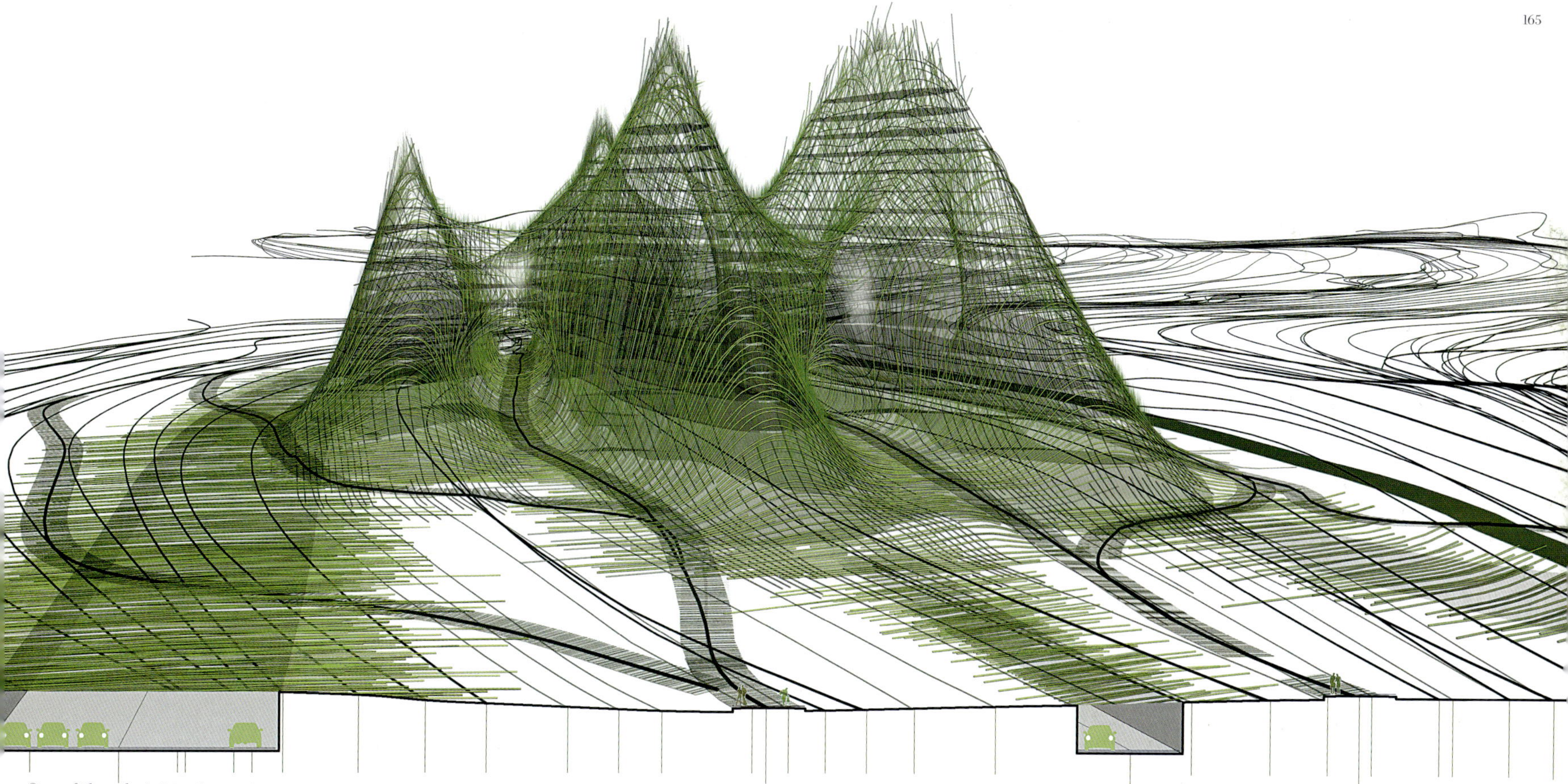

One of the administrative sectors:
The city plan is generated through geographic, programmatic, and environmental parameters.

ZXY scalar script:
Analysis of flow pressure in relation to the topography, with the preliminary organization of the volumes and profile of the city

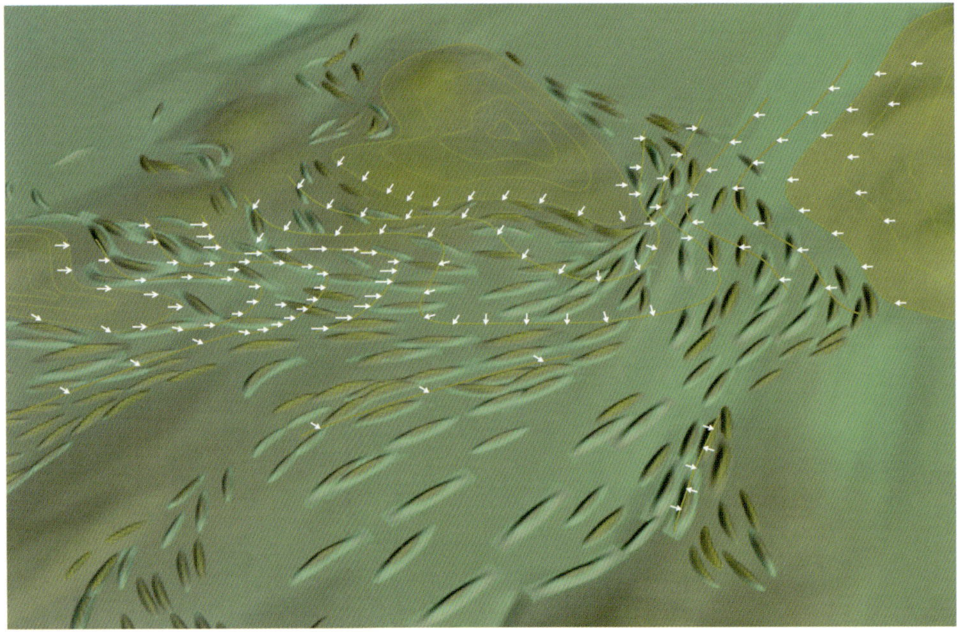

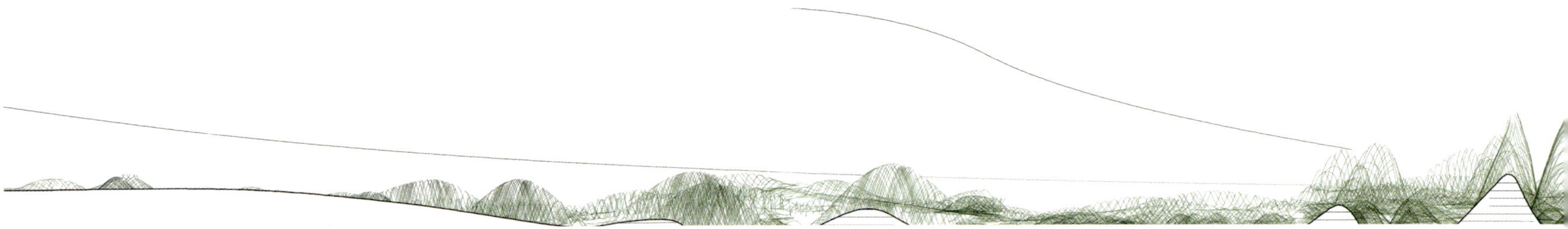

ZXY scalar script:
Example of treatment of the programmatic data regarding two usage categories: administration and residence with services

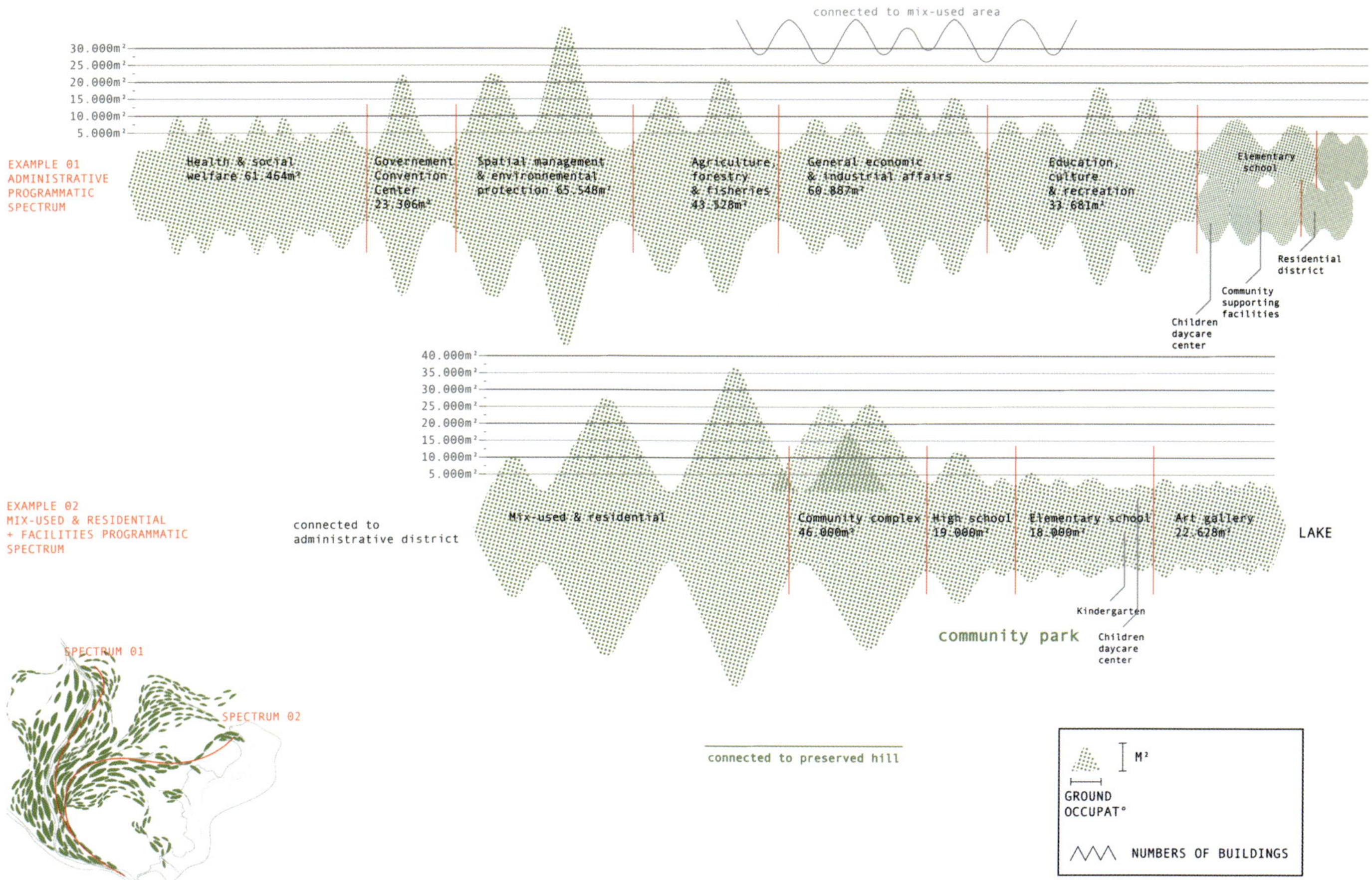

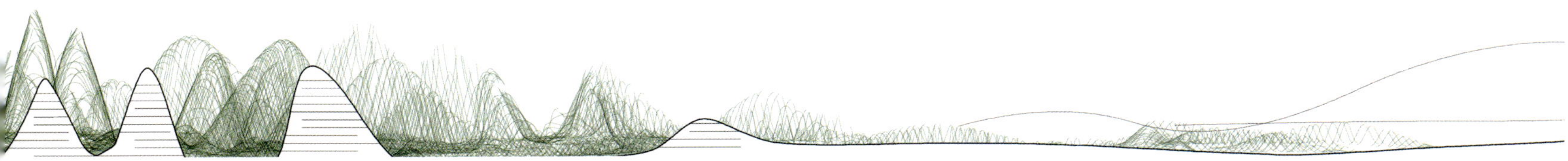

Nerves script:
The external skin with the robots; Swarm Town is based on a process of difference and repetition that also involves external-internal interfaces, in a context of adaptable and variable standardization.

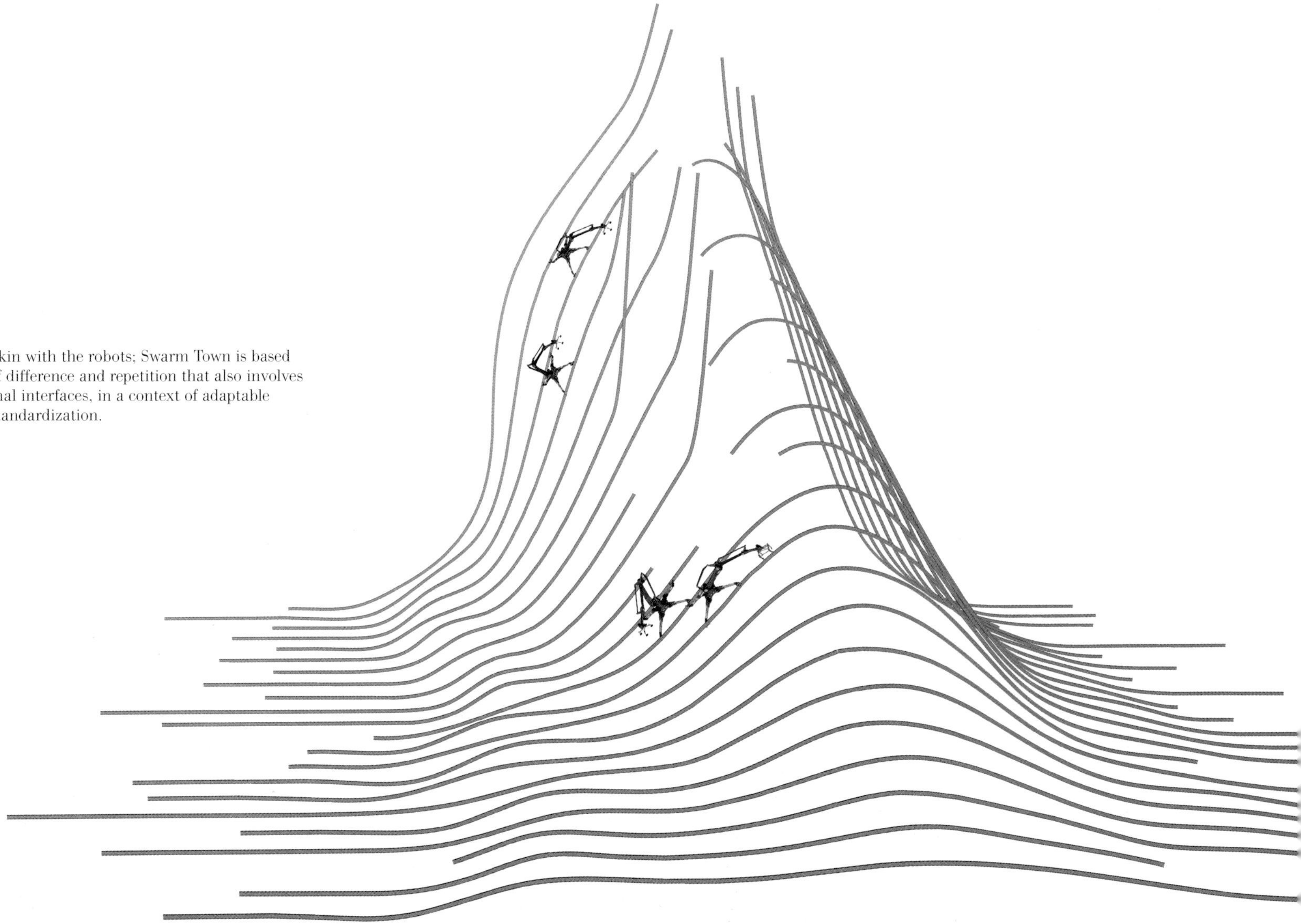

Nerves script:
Details of the robots, with the equipment for assembly of the structures and the facings

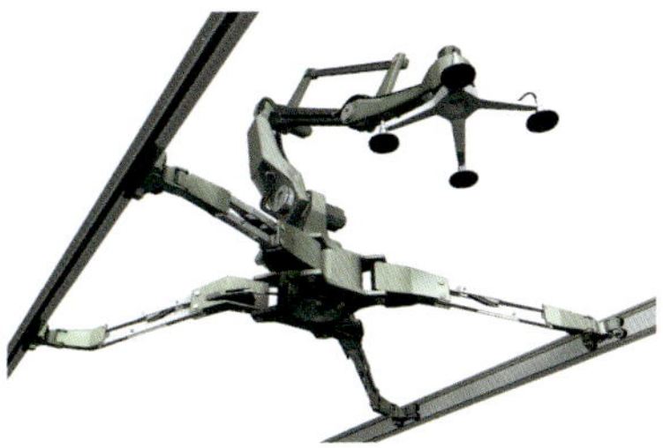

architects: R&Sie(n) year: 2008 location: Lopud, Croatia latitude: 42°41′ N average temperature: 59.9°F (15.5°C)

inflatable, public “restricted area” for an alchemical experiment with toxic plants—drugs, poisons, and antidotes

client: Francesca von Habsburg, TBA21 Vienna dimensions: 8,611 ft.^{2} (800 m^2)

delights and dangers of a garden of medicinal plants

Computer simulations:
The greenhouse descends like "toxic lava" on the slope of the terrain, in keeping with a gravitational protocol.

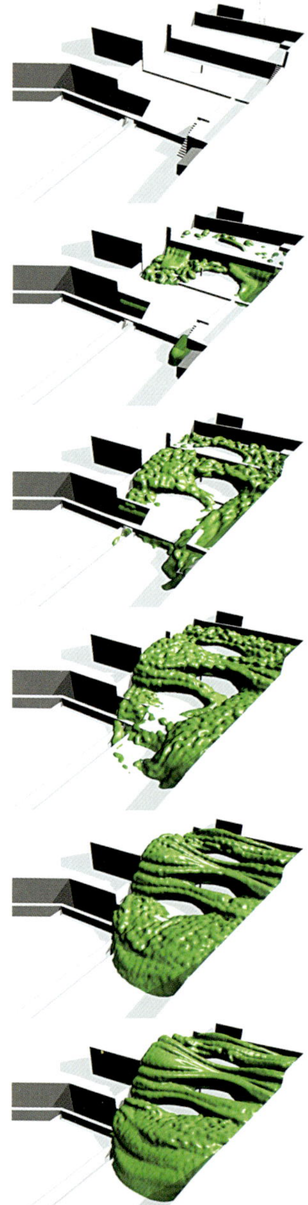

scenario

1)

Gathering of toxic plants for use as drugs, poisons, and antidotes: This tradition dates back to the Middle Ages in this Mediterranean environment (Franciscan monastery gardens).

2)

Emission of a toxic lava: The terraced morphology defines the new greenhouse, via a process of liquid viscosity.

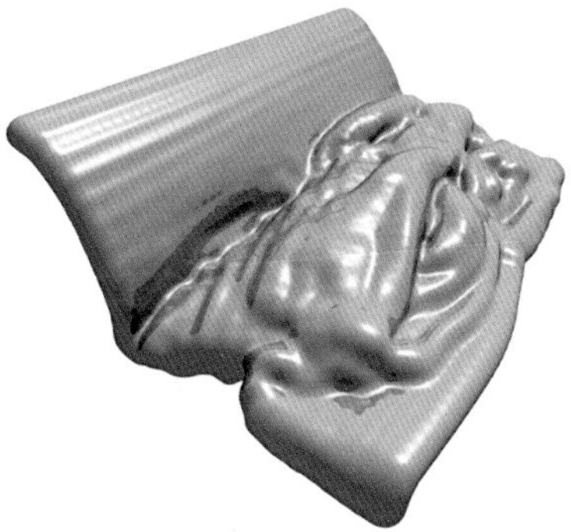

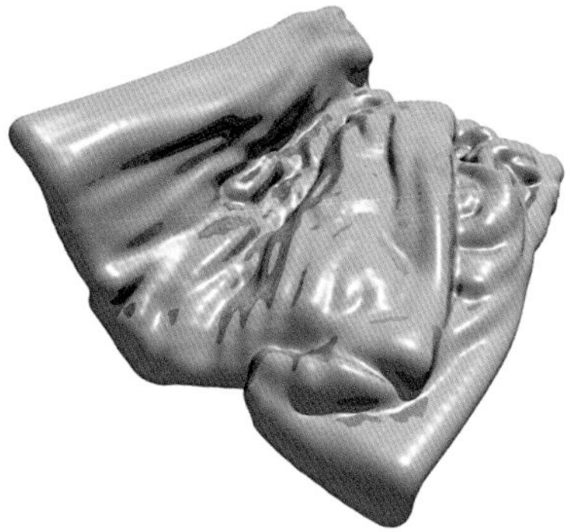

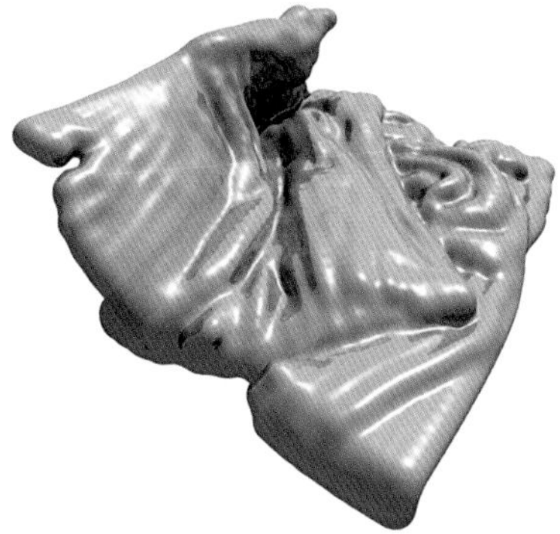

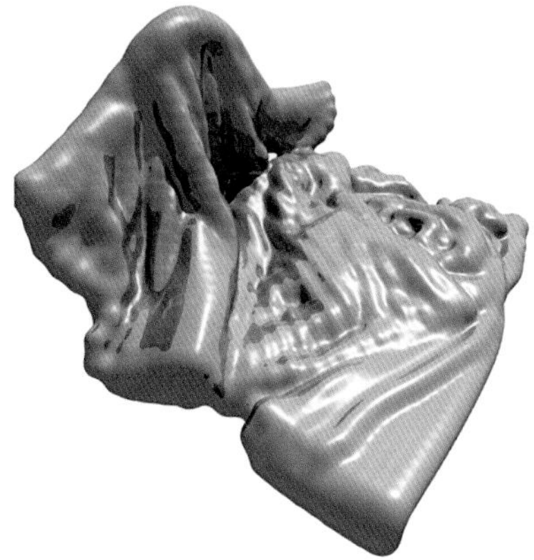

3)

Setting up of sharp spikes for supporting the fragile roof

4)

Definition of the restricted area for toxic cultivation

5)

Watering of the garden, thanks to the porosity of an inflatable, floating ETFE cushion

The project area, in the monastery of Lopud, and site plan

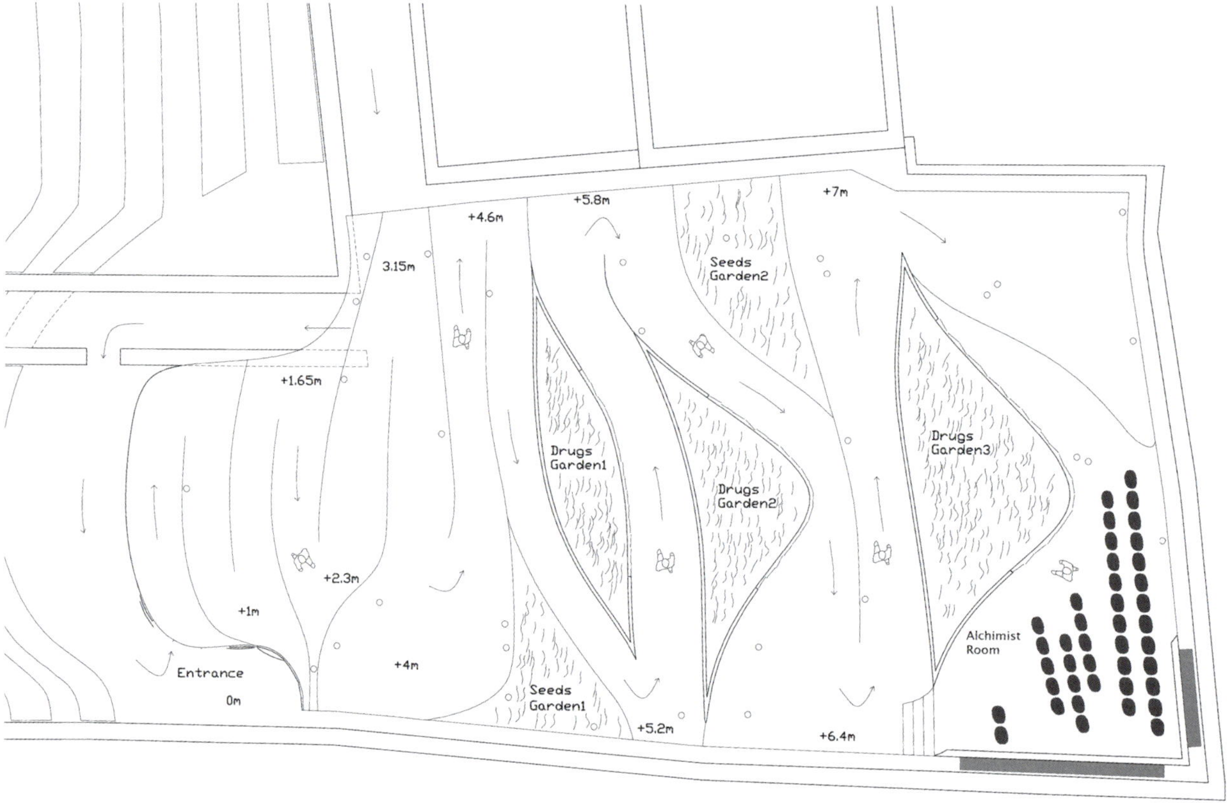

6)

Creation of a distillation machine for the metamorphosis of the plants into juices, ointments, and infusions

7)

Testing the of "Fugu"* effect of each plant on the metabolism, thanks to acquiescent visitors

*The Fugu, or puffer fish, is very famous in Japanese cuisine because its skin and internal organs are lethally poisonous and, though only specially licensed chefs should prepare it for the public, people die every year after having eaten it.

Plan with the organization of the three toxic gardens:
The higher the level, the more dangerous the plants are.

Low risk
1. *Adonis vernalis* (pheasant's eye)
2. *Lactuca virosa* (wild lettuce)
3. *Convallaria majalis* (Lily of the Valley)
4. *Paris quadrifolia* (Herb Paris)
5. *Mandragora officinarum* (mandrake)
6. *Asarum europaeum* (European wild ginger)

Medium risk
1. *Digitalis purpurea* (Common Foxglove)
2. *Helleborus niger* (Christmas rose)
3. *Datura Stramonium* (angel's trumpet)
4. *Dryopteris filix-mas* (male fern)
5. *Scopolia carniolica*
6. *Strychnos nux-vomica* (Strychnine tree)

High risk
1. *Colchicum* (autumn crocus)
2. *Ricinus communis* (castor oil plant)
3. *Daphne mezereum*
4. *Papaver somniferum* (Opium Poppy)

At the end of the ascent, a workshop for the preparation of remedies allows visitors to try herbal teas, infusions, tinctures, and other products.

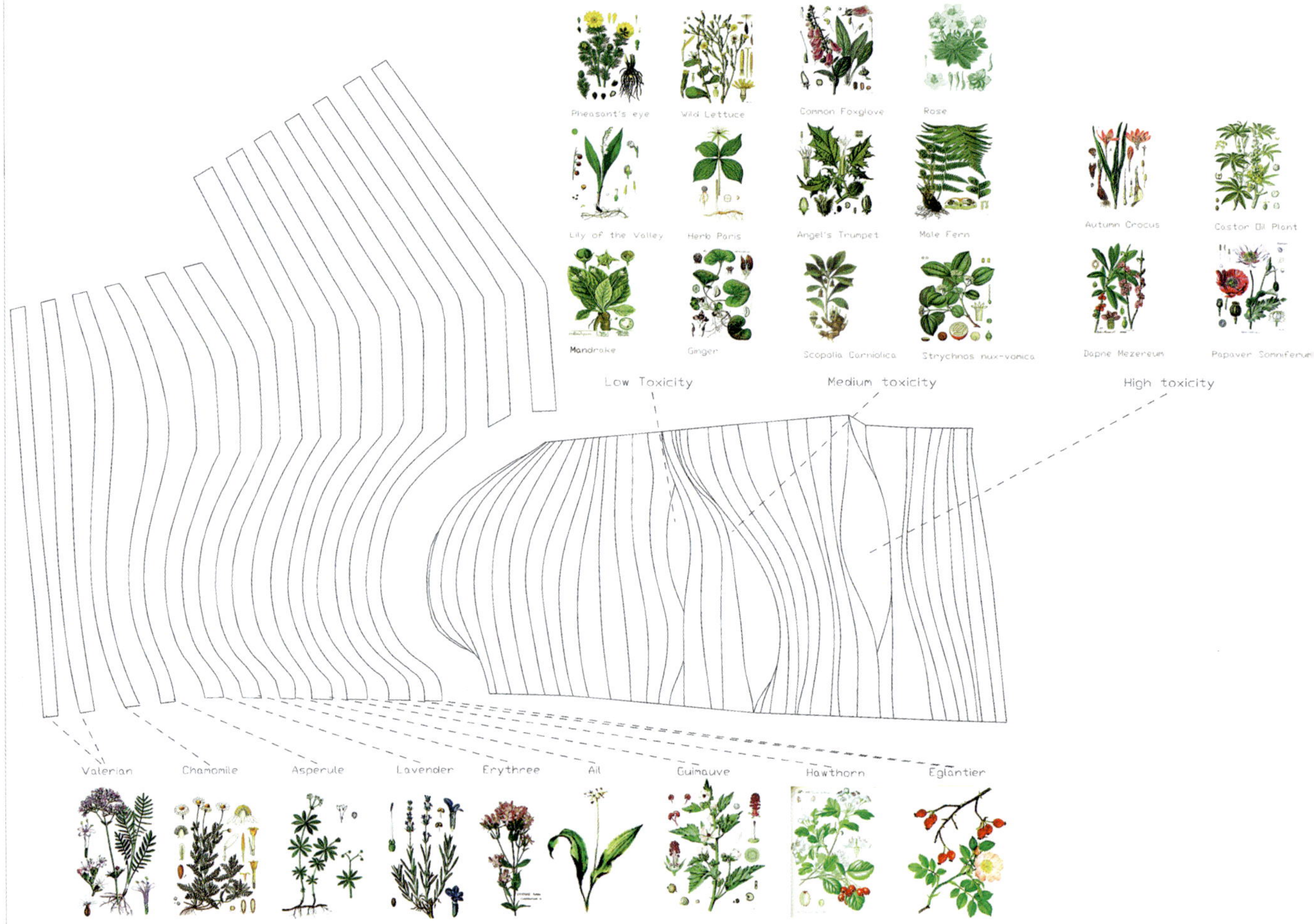

Interior views:
The transparent roof is made with inflatable cushions of ETFE plastic, capable of regulating the passage of light and the internal temperature. The many openings permit collection of rainwater for irrigation. Transparent walls protect the three toxic gardens, which can be opened to the public only under specific security conditions.

Devices and supports:
Each glass support holds three devices that can be connected to work in a series or in parallel. Each element makes it possible to remove the heat produced by processes of separation: desiccation and concentration.

The devices and their supports rotate to optimize operations. Each support is mounted on a steel pillar that guides the rainwater from the roof to the cisterns, to be filtered and input to the cooling circuits. The pillar also contains electrical wiring and pneumatic tubes to generate a vacuum where necessary.

architects: R&Sie... D/B:L   year: 2000   location: La Défense, Paris, France   latitude: 48°53'N   average temperature: 52.2°F (11.2°C)

renewable energy for electric and thermic production on an overexposed skin, burned by the sun and pocked by cancerous, degenerative mutations

client: Électricité de France (EDF)   cost: $60,000,000 (€45,718,000)   dimensions: 101,062 ft.^{2} (9,389 m^2)

hairy photovoltaic building with mutant skin

< (un)plug >

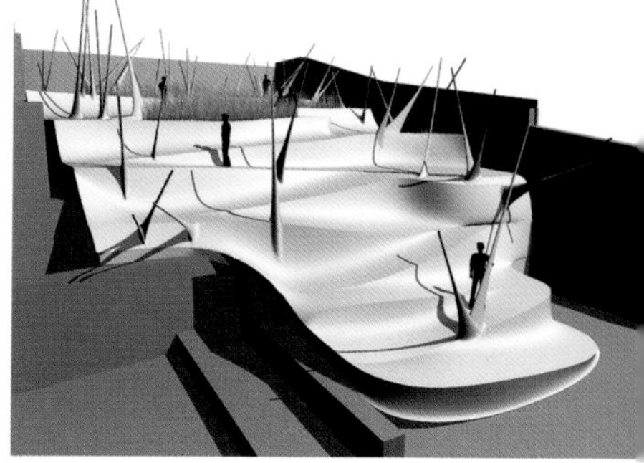

Section and model of the site topography:
The project, which is entirely reversible, responds to the need for conservation of the old monastery garden, with the aim of going beyond a mere reproduction of the past.

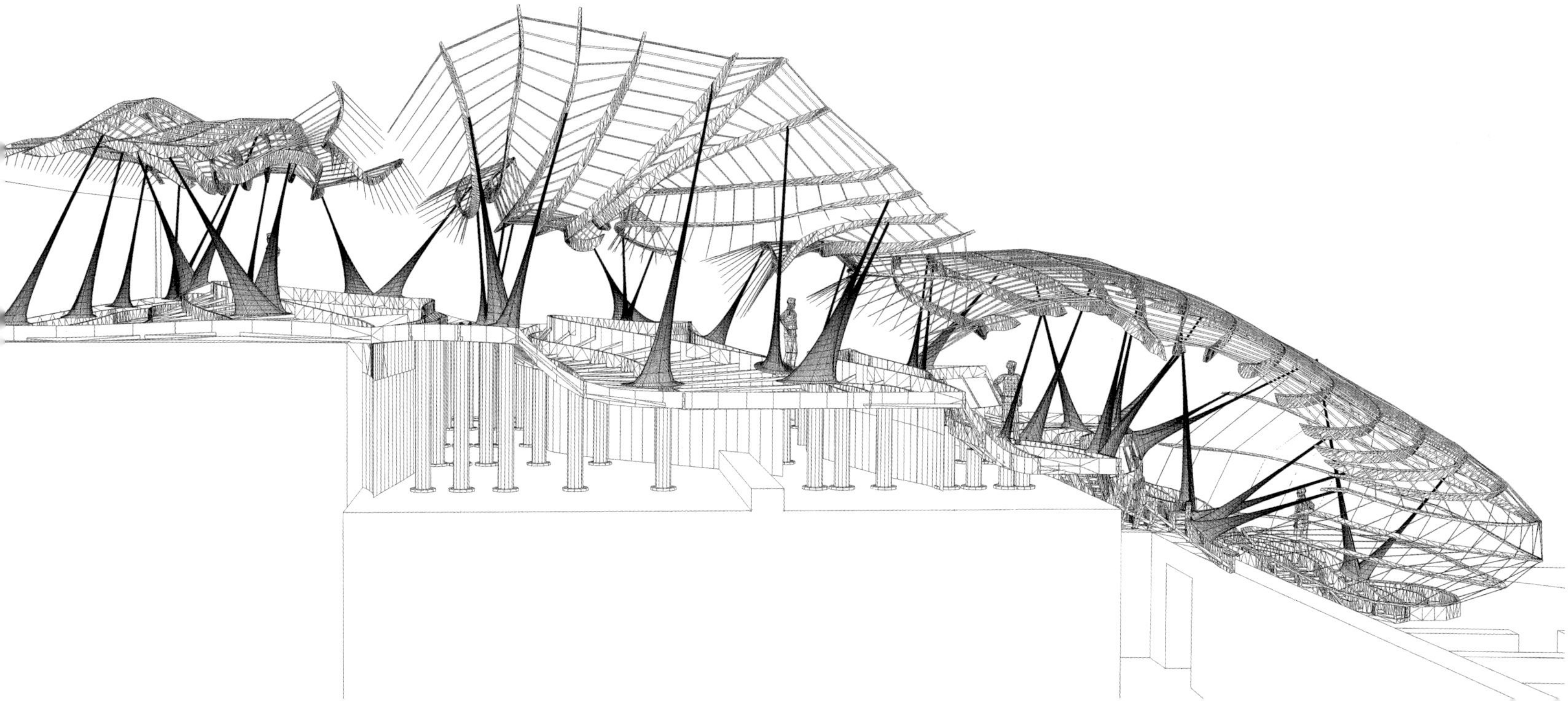

Detail of one of the photovoltaic outgrowths, conceptual model, and view of the project in Paris/La Défense

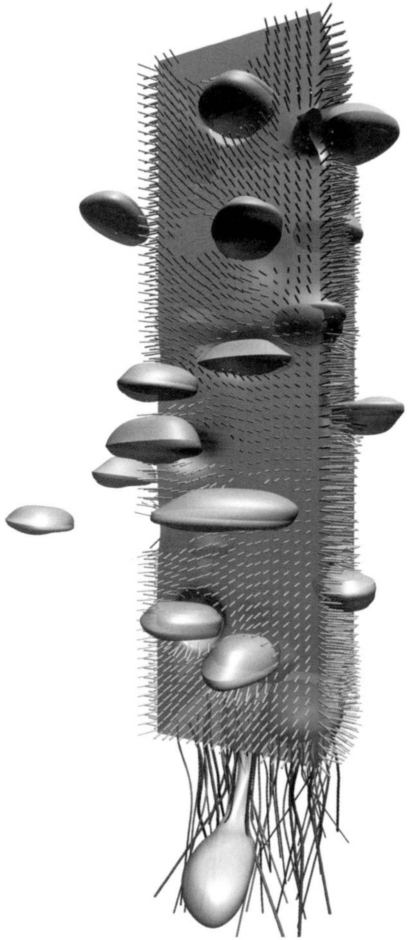

scenario

This is a “concept building,” similar to the concept cars automakers present at motor shows, with a reactive enclosure capable of producing renewable energy through solar vacuum tubes and flexible silicon panels.

1)

Construction of a standard office building

2)

Deformation of the reactive facade to gather renewable energies:
—"hairy" facade for thermal sensors (3 mi. / 5 km)
—"swollen" glass skin for photoelectric cells (21,500 ft.^{2} / 2,000 m^2)

3)

Disconnection (unplugging) of the building from the city power grid

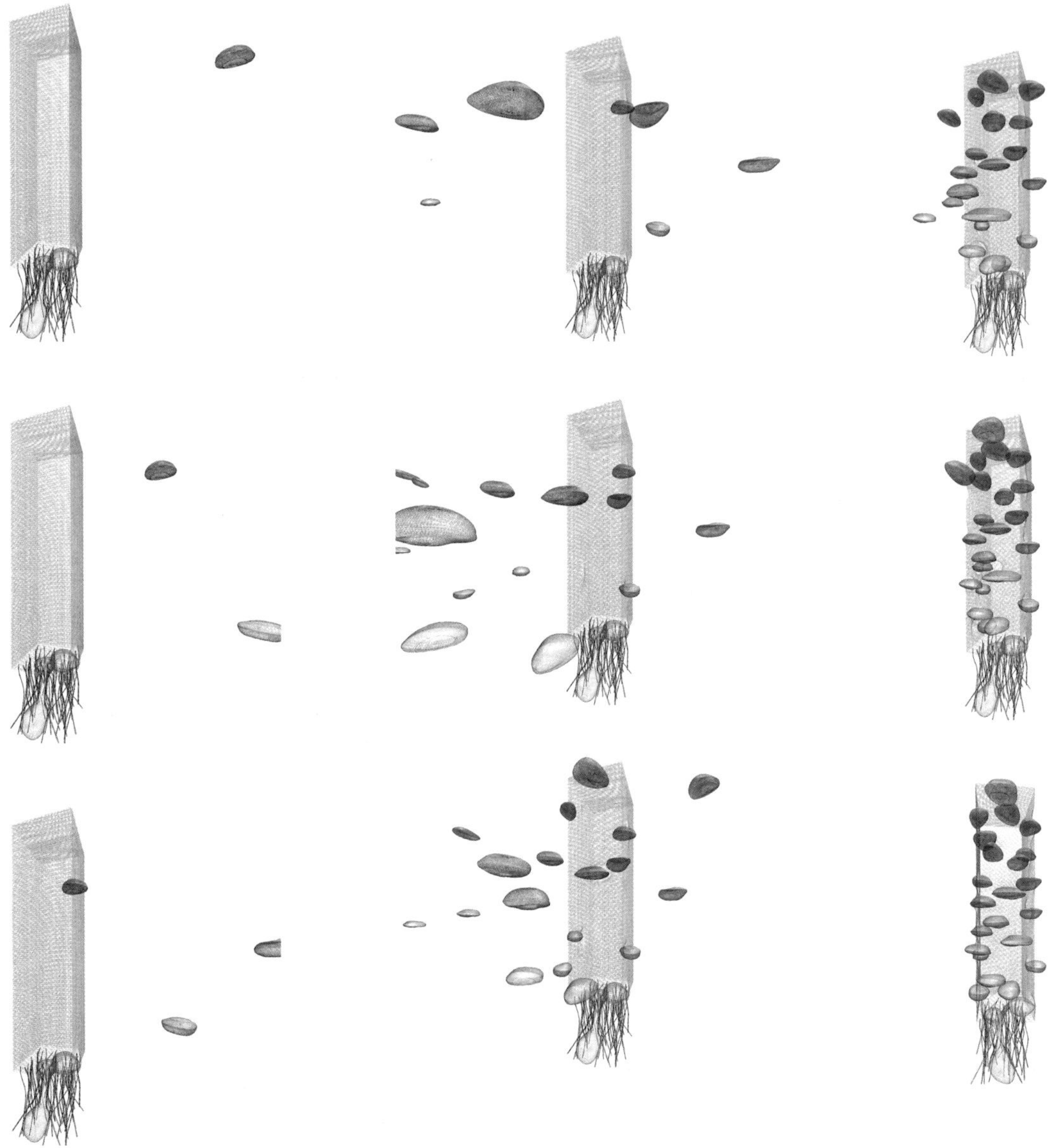

The thermal layer (with "hairs" composed of solar vacuum tubes), photovoltaic layer (the "outgrowths," covered with flexible silicon panels), perimeter structural grid, and enclosure

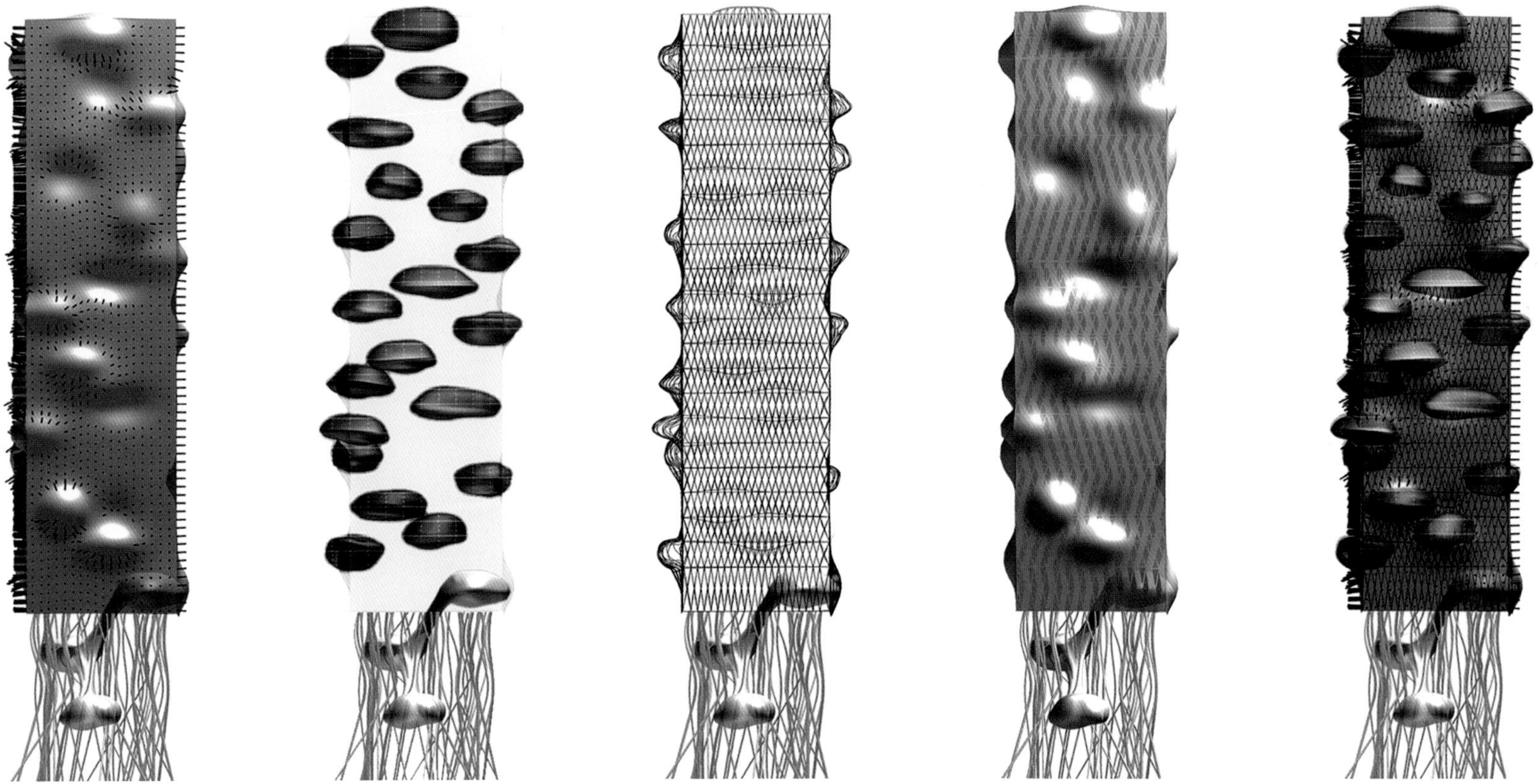

architects: R&Sie(n)   year: 2008   location: Paris, France   latitude: 48°52'N   average temperature: 52.2°F (11.2°C)

hydroponic, water-recycling, gaze-processing machine—a living envelope for a "rear window" game

client: confidential   cost: confidential   dimensions: 1,400 ft.^{2} (130 m^2)

home of the architect, with a hydroponically nourished plant covering

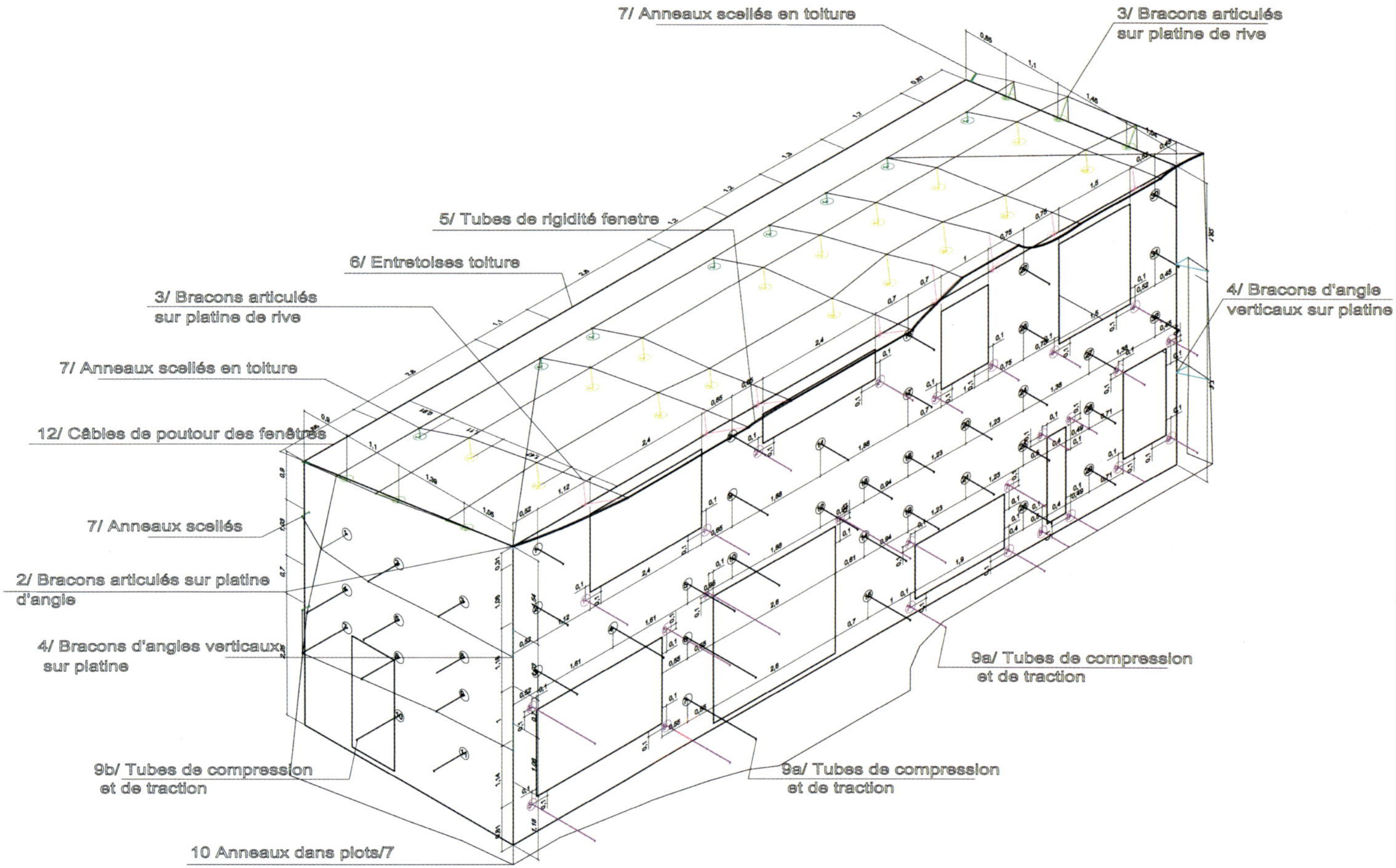

scenario

1)

Design of a private laboratory, in the style of a hunter's "duck blind" shelter

2)

Creation of an envelope made from twelve hundred hydroponic ferns

Structure of the hydroponic garden, and fabrication of the phials for the nourishment of the plants

3)

Introduction of three hundred glass beakers: These provide extra light for the interior, through refraction. They also allow for a *Rear Window* (as in the Hitchcock movie) negotiation of the closed courtyard.

Details of the fern garden's support structure and view of the work site

③ Bracon articulé sur platine de rive/14

N	longueur	quantité
1	1120mm	2
2	780mm	2
3	1030mm	2
4	800mm	2
5	362mm	3
6	452mm	1
7	542mm	1
8	633mm	1

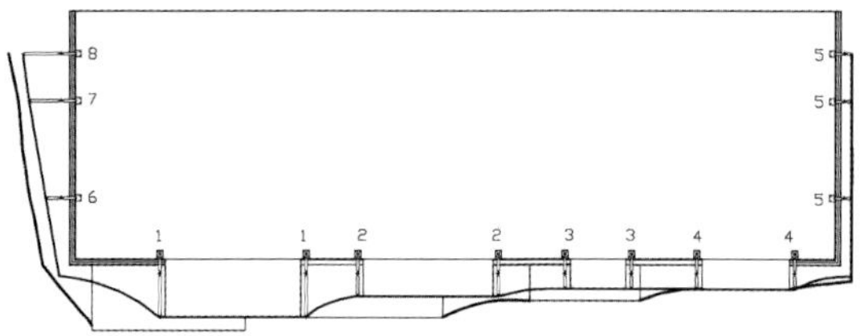

Plane Esc 1:100

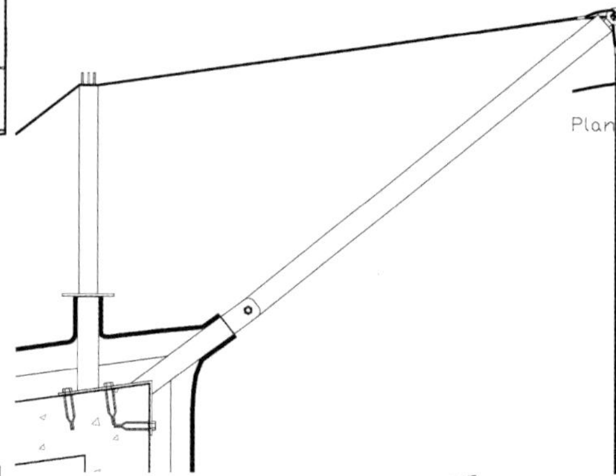

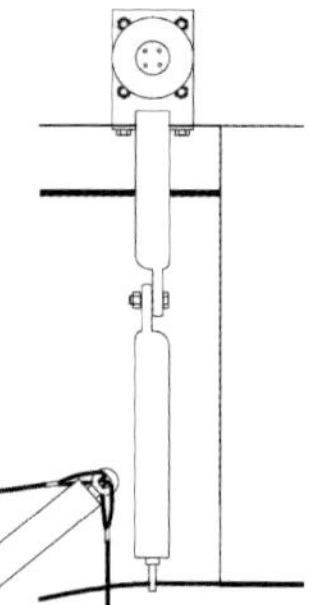

Plane Esc 1:10

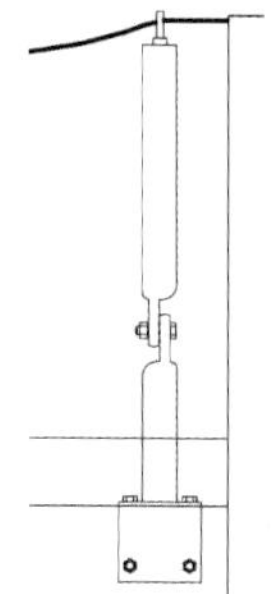

Façade Esc 1

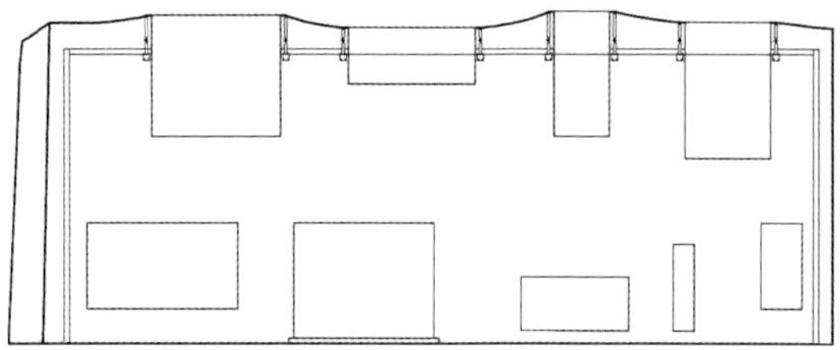

Façade Esc 1:100

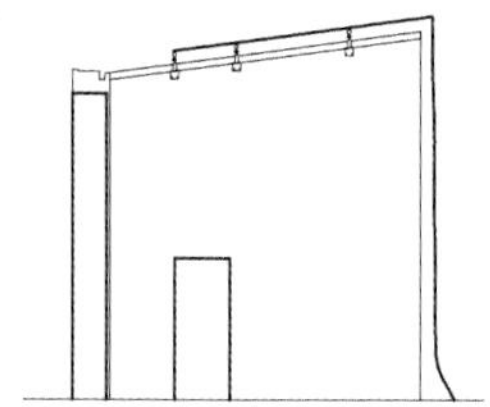

Façade Esc 1:100

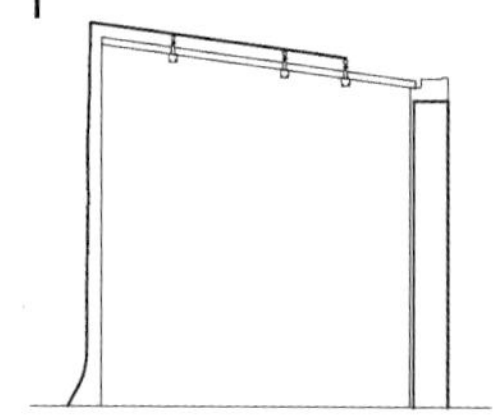

Façade Esc 1:100

4)

Collection of rainwater, in order to water the plants: This is done via an individual mechanical-drip system. The drip system also provides nutritional additives and portion controls.

5)

Formation of "Devil's Rock"-like protuberances (as in the strange mountain in Spielberg's *Close Encounters of the Third Kind*) with a prehistoric nature: The ferns are from the Devonian period. These are technologically domesticated in order to come back in the current "regressive" French period.

Positioned at the windows, the hydroponic phials, left without vegetation, refract light inward and block the view from outside.

Arrangement of the hydroponic phials
on the three free facades and the roof

	longueur	quantité
Façade	30*30mm	486
Pignon nord	30*30mm	154
Pignon sud	30*30mm	41
Toiture	30*30mm	308
	Total	989

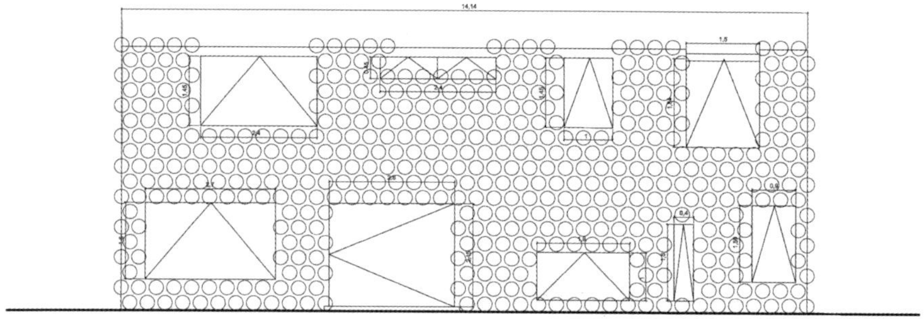

Façade Ech 1:100

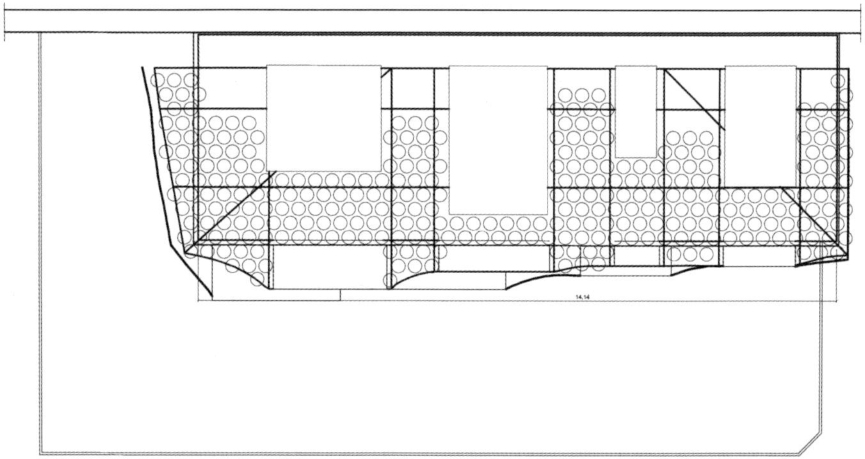

Toiture Ech 1:100

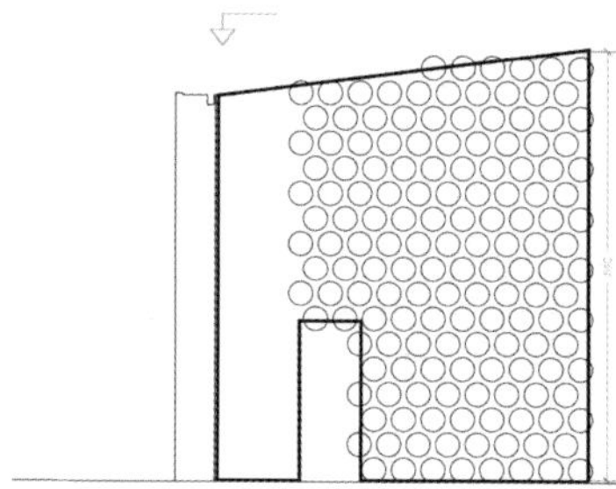

Pignon Nord Ech 1:100

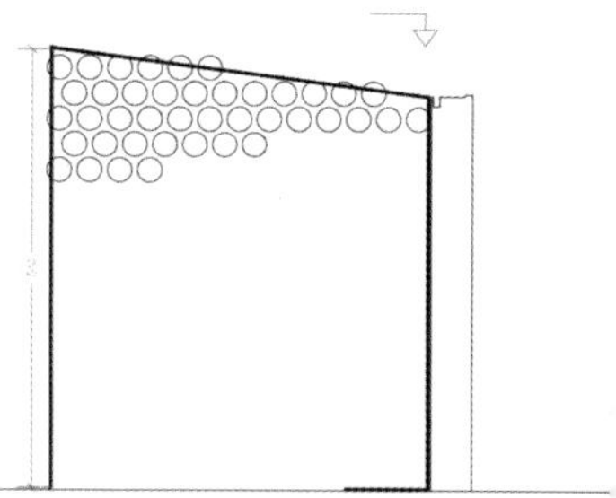

Pignon Sud Ech 1:100

Views of the interior of the interstitial space and the vertical garden:
The concrete walls are covered with insulating material and geofabric.

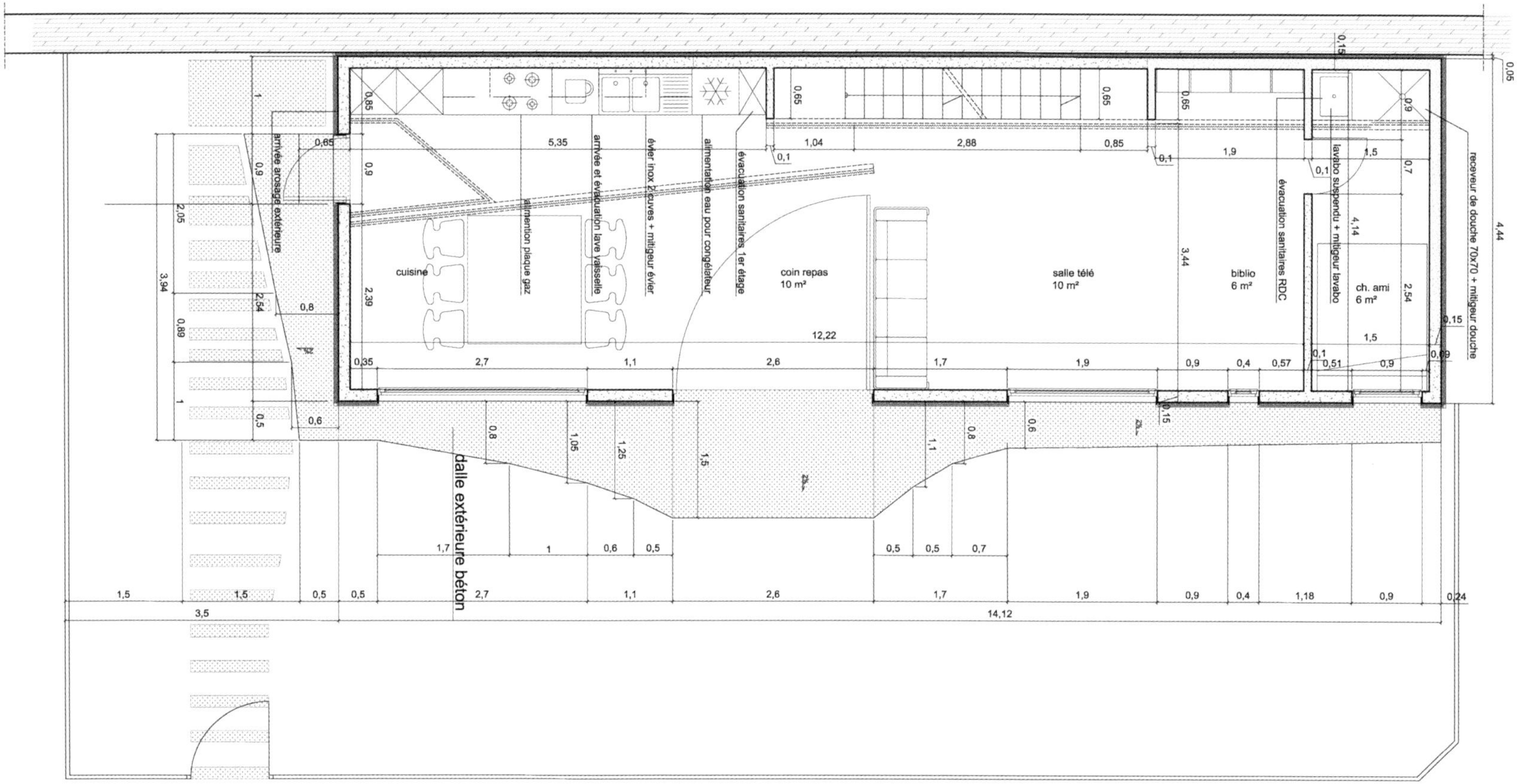
cuisine
coin repas
10 m²
salle télé
10 m²
biblio
6 m²
ch. ami
6 m²
arrivée arosage extérieure
alimentation plaque gaz
arrivée et évacuation lave vaisselle
évier inox 2 cuves + mitigeur évier
alimentation eau pour congélateur
évacuation sanitaires 1er étage
évacuation sanitaires RDC
lavabo suspendu + mitigeur lavabo
receveur de douche 70x70 + mitigeur douche
dalle extérieure béton

The ground floor plan (opposite) contains the dining room, kitchen, and living room, while the upper level (below) has the bedrooms of the parents and the two children. On the following pages are the dining room and the living area.

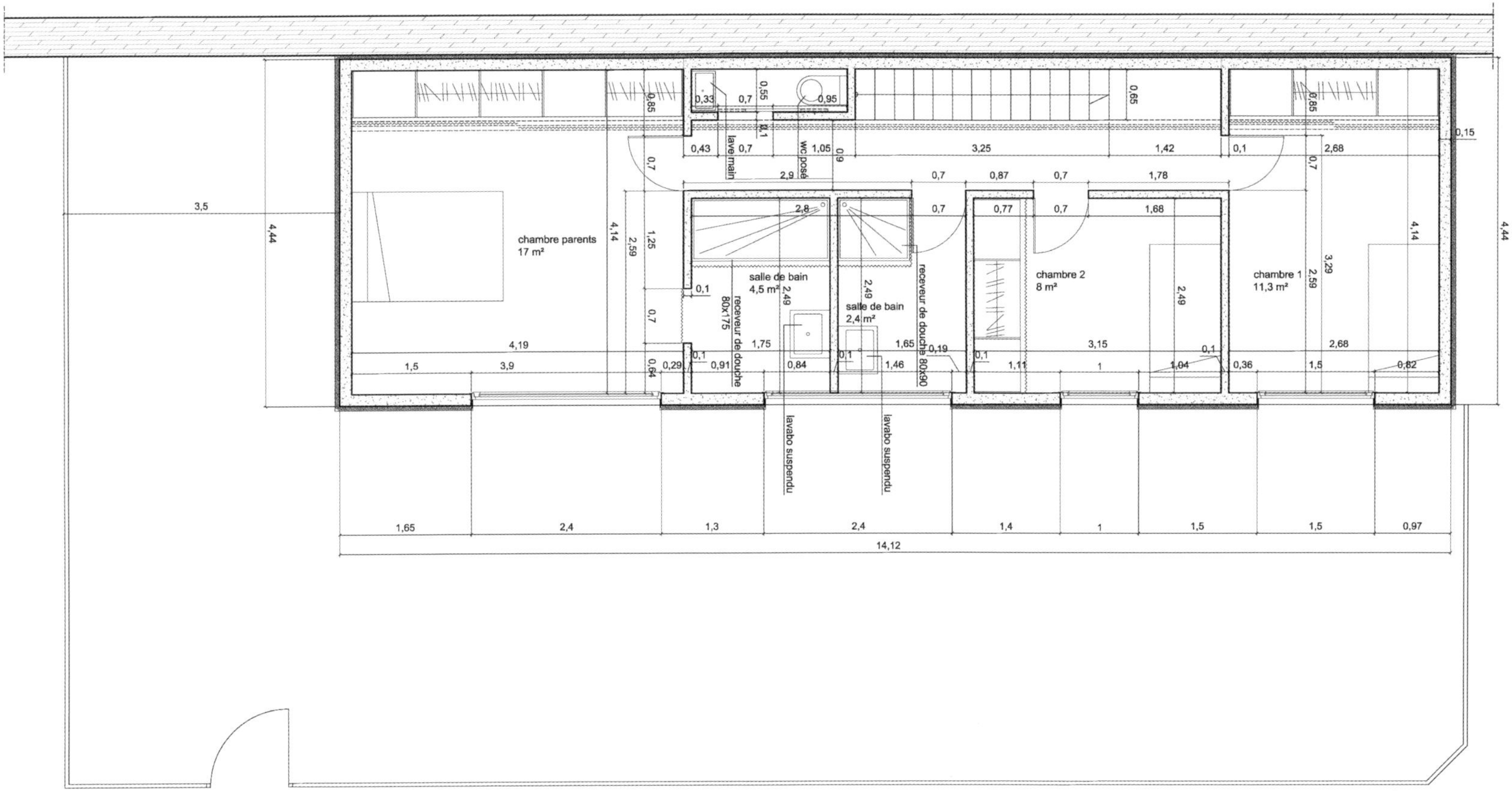

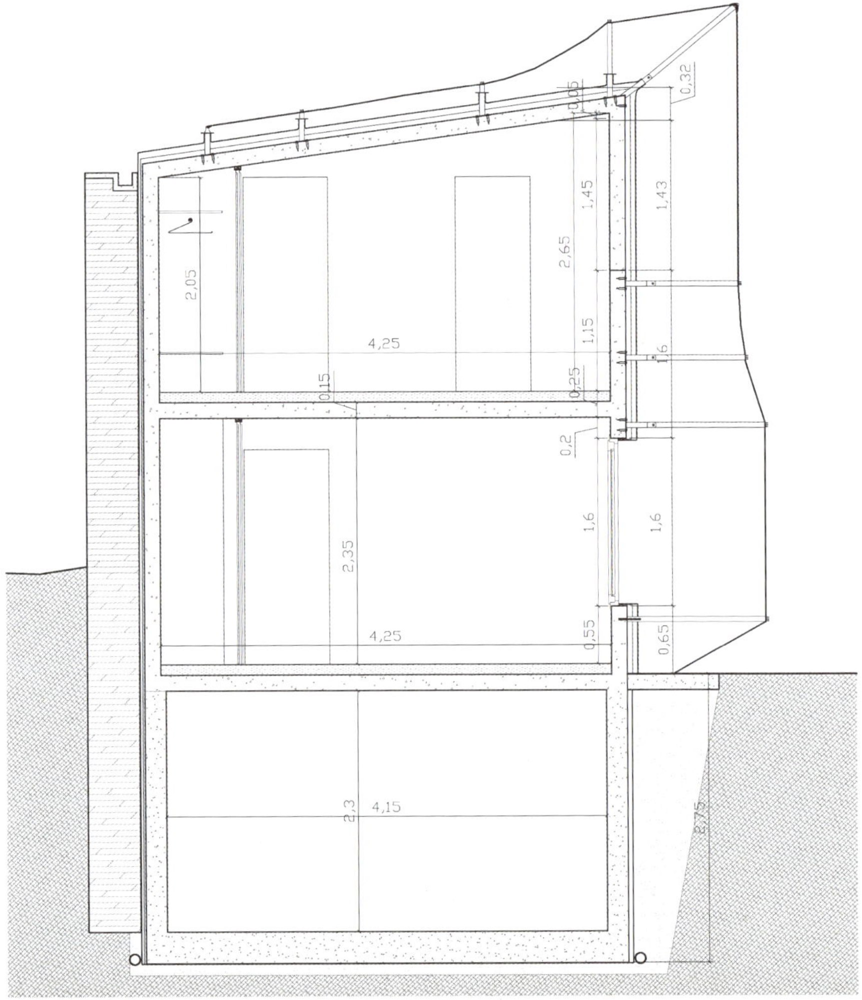

Cross section and views of the growing garden

multiple identity

1. The studio's letterhead has the name R&Sien Architects at the upper right; below, it becomes R, D & Sie . D / B : L. At the start of his architectural career, in 1990, François Roche signed his projects with his full name, adding the names of the others who had taken part in the various initiatives. From 1993 to 1999 almost all the projects were signed Roche, DSV & Sie. For a more complete chronology, see the credits section.

2. R&Sie sounds like R. & Cie., but there is also a clear pun with the word *hérésie.* (heresy).

3. "Unfinished Business," François Roche, interviewed by Jeffrey Inaba and Benedict Clouette, *Volume* 10 (2006): 26–30. In spite of the reluctance to have a public image in the media, François Roche appears in a couple of the images that accompany his firm's projects and constructions (images not included in this book): looking for them can be a bit like watching the films of Alfred Hitchcock, waiting for the director's inevitable cameo.

4. See, among the many online sources, the entry at Wikipedia: http://en.wikipedia.org/wiki/Luther_Blissett_%28nom_de_plume%29.

utopia

5. The Paranoid-Critical Method is an intuition of Dalí revived by Rem Koolhaas in *Delirious New York* (London: Thames & Hudson, 1978): "As the name suggests, Dalí's Paranoid-Critical Method is a sequence of two consecutive but discrete operations: 1. the synthetic reproduction of the paranoiac's way of seeing the world in a new light—with its rich harvest of unsuspected correspondences, analogies and patterns; and 2. the compression of these gaseous speculations to a critical point where they achieve the density of fact," 202; "The PCM . . . promises that, through conceptual recycling, the worn, consumed contents of the world can be recharged or enriched like uranium. . . . The PCM proposes to destroy, or at least upset, the definitive catalogue, to short-circuit all existing categorizations, to make a fresh start—as if the world can be reshuffled like a pack of cards whose original sequence is a disappointment," 203.

fear and loathing

6. D'Arcy Wentworth Thompson, *On Growth and Form* (Cambridge: Cambridge University Press, 1917).

disorientation

7. Eyal Weizman, "Lethal Theory," *Log* 7 (2006). Weizman's article describes how the Israel Defense Forces (IDF) developed and deployed war tactics in the Palestinian territories derived from the thinking of Debord, Deleuze, and Guattari. "This form of movement, described by the military as 'infestation,' sought to redefine inside as outside, and domestic interiors as thoroughfares. Rather than submit to the authority of conventional spatial boundaries and logic, movement became constitutive of space. The three-dimensional progression through walls, ceilings, and floors across the urban bulk reinterpreted, short-circuited and recomposed both architectural and urban syntax. The IDF's strategy of 'walking through walls' involved a conception of the city as not just the site, but the very medium of warfare—a flexible, almost liquid medium that is forever contingent and in flux." 53.

8. Jorge Luis Borges, *El jardín de senderos que se bifurcan* (Buenos Aires: Ediciones Sur, 1942).

base materialism

9. For more on Bataille and on the cultural and artistic implications of refuse, decay, and entropy in general, see Yve-Alain Bois

and Rosalind Krauss, *Formless: A User's Guide* (New York: Zone Books, 1997).

10. "Adolf Loos and I—he literally and I linguistically—have done nothing more than show that there is a distinction between an urn and a chamber pot and that it is this distinction above all that provides culture with running room. The others, the positive ones, are divided into those who use the urn as a chamber pot and those who use the chamber pot as an urn." Karl Kraus, *Die Fackel* (1912), 37, cited in Hal Foster, *Design and Crime* (London: Verso, 2003), 16.

11. "Here 'those who use the urn as a chamber pot' are Art Nouveau designers who want to infuse art (the urn) into the utilitarian object (the chamber pot). Those who do the reverse are functionalist modernists who want to elevate the utilitarian object into art. (A few years later Marcel Duchamp would trump both sides with his dysfunctional urinal, *Fountain*, presented as art, but that's another story.)" Ibid., 17.

danger

12. In his famous essay "Das unheimlich" (1919), Sigmund Freud writes: "*Heimlich* is a word the meaning of which develops in the direction of ambivalence, until it finally coincides with its opposite, *unheimlich*." Cited in Anthony Vidler, *The Architectural Uncanny: Essays in the Modern Unhomely* (Cambridge, MA: MIT Press, 1992), 25.

13. "At once a psychological and an aesthetic phenomenon, [the uncanny] both established and destabilized at the same time. Its effects were guaranteed by an original authenticity, a first burial, and made all the more potent by virtue of a return that, in civilization, was in a real sense out of place." Ibid., 27.

machines and narratives

14. http://www.thelandfoundation.org/?About_the_land.

15. "Astonishingly, (science) fiction has shifted neither forward nor rearward but into the here and now. The unfolding scenarios it follows to manipulate our reality are becoming true transformation tools and paradoxically strategic levers to grasp the wobbling of our postdigital societies, our choked mass media culture." François Roche, (Science) Fiction & Mass Culture Crisis, in R&Sie... Architects, *Spoiled Climate*, ed. Andreas Ruby and Benoît Durandin (Basel: Birkhäuser, 2004), 56.

toxic ecologies

16. David Usborne, "Toxic waste creates hermaphrodite arctic polar bears," the *Independent*, 10 January 2006.

17. In the film *Charisma*, directed by Kiyoshi Kurosawa in 1999, a prehistoric tree produces a toxin that poisons the surrounding forest in a zone used for lumber. The plot is based on the violent conflict between two different ideas of conservation of nature.

endlessness

less...

chat

endlessnessless...

with
giovanni corbellini
benoît durandin
joker
alexandra midal
françois roche

Endlessnessless... is an email conversation that took place from October to December 2007. Initially imagined as a sort of ship's log of the book, or a "making of" document to record its hypotheses and changes of direction, it became an autonomous machine, running with its own rules. François Roche, the inner voice of R&Sie(n) and "master of ceremonies" of this chat, called on Giovanni Corbellini, an architecture critic, Benoît Durandin, a scholar of philosophy of science, and Alexandra Midal, an expert on media culture and "celibate machines," to take part in a sort of role playing game, where each would contribute a specific viewpoint. All the participants in the conversation then played, from time to time, the role of a fifth, fictional character, the Joker, whose unpleasant, cynical character made it possible to anonymously advance criticisms, doubts, and more-or-less-inconvenient questions. This chat is an "excellent cadaver" mechanism capable of immediately bringing out the obsessions, key words, problem nodes, cultural references, and narrative strategies of a complex, transgressive research.

In a multiple personality disorder (in fact in my own multiple personality disorder), the only way to define theories and inspirations is to create an incest between them, to escape-interpret-speculate...

...from within and beyond mass-media culture. I have bought a tube of red lipstick to trace my reflection on the mirror, with nine apparatus, to define how strategies of relationships articulate knowledge and are simultaneously articulated by themselves.

f. r.

Giovanni Corbellini
What's this, green monsters? [looking at image on p. 205]

François Roche
Which monsters, did you see monsters?

Giovanni
I think so (but if I look at them a little bit longer they begin to look funny).

Alexandra Midal

you as he runs away. I'd rather go for Snow White mixed with girly Victorian fairy tales. Girls are not paralyzed by the proximity of danger in the forest. In addition, don't you feel a slight and exciting insinuation of casual sadism? I do!

Benoît
I feel just like time is freezing, an impression of being sucked up. Look at their paws! They are leaving trails

all, for it runs through everything."[5] Look at the trajectory: it's so erratic, full of noises, searching for the next movement, winding and winding and winding, writing on itself. Writing against architecture.[6] Maybe that which we have woken up is the acephalous man, made of bones, guts, and nerves: mesoderm, endoderm, and ectoderm all folded together.[7]

François

It's from the paranoia of the two little girls, who are about to perform a productive and not-at-all-innocent routine.

Benoît Durandin
You mean you've never seen that ritual? Like the reproduction of something on itself, out of its matrix?

Alexandra
No, as a little girl myself, I saw them, I saw them [humming]: "Sometimes I'd divide, / And burn in many places; on the top-mast, / the yards and borespirit, would I flame distinctly."[1]

Joker
With this kind of stargate of a parallel universe, are you sure it's a book about architecture?

Giovanni
Who said that?

Alexandra
What?

Benoît
So this is a book about architecture?

François
Nobody said the contrary.

Alexandra
And why not?

Giovanni
So the paranoia is the key?

François
More than a key, it's an "apparatus."[2]

Giovanni
If we suppose now that these Hulks should be real…

Alexandra
But don't threaten them, they could be like the white rabbit.

Giovanni
The one of Alice's?

Alexandra
Don't you know? The rabbit hypnotizes >

everywhere, bloody and green fluorescent footsteps, indicating I don't know what…a passage maybe.[3] It looks now like letters.[4] I don't know if…

Joker
You fools, all I can see is a hair ball, pushed by the wind.

Benoît
Yes, but it's pushed by Aeolos, the god of winds. "The mind is the quickest of >

In a paranoid sense, it's like an apparatus, submitting and being submitted by a large spectrum of interpretation, from self-illusion to freak 3-D effects. We could consider this "moment" as the first apparatus, or, more precisely, as how strategies of relationship, embedded in the moment, articulate knowledges that are simultaneously articulated by themselves. The two little girls are looking at the product of their own paranoia, in this way transforming at the >

apparatus 1: paranoia of the little girls

It is an apparatus of illusion and reality, where the projections of the mind are enough to create green monsters. Like Alice in Wonderland, combining fiction with a touchable universe, we are now definitively plugged into this schizoid and paranoid biotope from the mass-media culture crisis. When Marcel Duchamp discovered Xenakis's music and Philips Pavilion (and, accessorily, Le Corbusier), he hypothesized that these productions were infiltrated and contaminated equally by mathematics and little girls. Behind this provocation was something about the subjectivities articulating the theory of correspondence between music and space, for example, the UPIC software of Xenakis. The fiction became the most important key to our alienation, with or without the little girl.

f. r.

same time the representation of our own reality. They articulate fiction and reality by creating, through their subjectivities, a narration that renegotiates with touchable substances. We are directly confronted by the boundaries of the system, where, from this scripted confusion, emerges an apparatus that at the same time reveals the boundarylessness of self-consciousness.

Alexandra
Not unlike déjà vu: the situation where you are convinced you have already experienceed an identical thing at another time—if you don't take seriously the rationalist approach, of what scientists describe as a breach or alteration between the brain's two hemispheres, a synchronization that therefore creates a shift of perception. And if you buy the idea of an inexplicable collision between many parallel universes, like Henri Bergson—who dedicated a pertinent essay to the question with "Le Souvenir du présent et la fausse reconnaissance"—a weakening of "the function of reality" is elucidated.[8] You get a momentary unfolding of the person, something like the famous burlesque paradox set up by the physicist Erwin Schrödinger in 1937. He thought up an experiment in which a cat, a radioactive particle, and a mechanism made up of a Geiger counter, a hammer, and a vial of lethal gas were locked in a closed box. If the atom disintegrated during a given amount of time, the counter would be activated, the hammer would break the vial, gas would escape from the vial, and the cat would die. In a space ruled by the laws of classical physics, there is as much probability that the atom would disintegrate as that it would not. According to the laws of quantum physics, two possibilities superimpose one another; the atom is at the same time activated and disintegrated; the cat is subjected to a state of indecision, at the same time dead and alive. This simultaneity is only accomplished in the instant when an outside individual observes the interior of the box. This principle, which dominates the subatomic and unknown world of our request of deterministic certitudes (in terms of cost, time, performance, security) has more and more to deal with an increasing instability of programs, tasks, and opportunities. So, the observer (which can be interpreted as the external conditions) becomes the main character: the one who unintentionally decides if the cat will live or die. Quantum physics teaches us that we can manage this interactive relationship in only a very paradoxical way. Should we crossbreed Schrödinger's cat with Deng's (which catches mice independent from whether it is black or white)?[10]

Alexandra
Hey! Joker, maybe you are just a BInary digiT, not even a bug! Bzzzz! Who cares about grasping the split between two worlds that even physicists can't explain? Rather, I think it is a basis to shift our egocentric and regular viewpoint and start thinking in a cynical fashion, as Flaubert brought up with *Le Dictionnaire des idées reçues*, in *Bouvard et Pécuchet*, or naively, as Brian Aldiss and his novel *Cryptozoic* claim.[11,12] But yes, I'll eventually go for some laboratory cat's experiments. What in hell would we do without cats? They seem to be crucial for the Western world's way of thinking!

Giovanni
You are right. Shifting the viewpoint is our main goal. As designers, we dreamt of that power that we fight as citizens. So the schizoid situation between planning needs and unpredictable developments to walk through walls.[15] Science and architecture share the same ambiguous and irrefutable relationship to reality. And this relationship creates frictions with unexpected results. It's with those unexpected results that we have to deal, not as prophets (too comfortable) nor outsiders (too reassuring) nor experts (too romantic)...

Joker
Schizophrenia, paranoia, cats, rabbits, mice... Are you undergoing pet therapy?

Giovanni
Good idea! We could use animals to feed our personality disorders. Laboratory hybrids or "natural" mutations are both fine. François often talks about hermaphrodite polar bears...

François
So you know the story, again and again. It is about the archipelago of Svalbard on the north pole, right above Norway.[16] This territory seems to be virgin, the last spot untouched by global pollution. But it is exactly the opposite: two toxic streams are leeching the western and eastern coasts. One is an arctic stream carrying radioactive substances from the Murmansk Sea and, more precisely, the corpses of Russian nuclear

apparatus 2: 3-D effects from entertainment to freakiness

What a hell exists in digital design, from keyboard addiction and computing as a voluntary loss of identity and integrity, a loss of paradise (with Milton in the background). In one case the effect just justifies the skill of the nerd, and in the other it rewrites the

of two or more worlds simultaneously, so-called parallel worlds. No future, no present, no reality? "You can pick up the world!" could be the White Rabbit's motto.

Benoît

It sounds like a Spinozian motto, "The mind endeavours to conceive only such things as assert its power of activity."[9] It always comes from and leads to experience as the only thing that we could not erase: fictions, whatever they are made of. Quantum physicians used to explain quanta theories through short stories, as fairy tales in a way, to transpose what they saw into something else less astonishing. "Four fish are swimming in a pool, two floodgates open simultaneously for two other pools, and at the end four fish are swimming in the two new pools." It's coherent in quantum logic, pertinent in a field of research, and uses cognitive accessibility: the three tensions required for any scientific knowledge. It also reveals the contortions that scientists have to do to reach a level of understanding when facing quantum properties. They have to bring noncrazy hypotheses (or theories) into the field of common knowledge, but when doing that, those noncrazy hypotheses become vectors of transformation.

Joker

Quantum physics is about ultrasmall scale. What happens to subatomic particles does not have much to do with the actual world where we live. I hope that the engineer who designed the bridge I cross every morning did it in a very deterministic way...

Giovanni

Maybe it is better to hope that our universe is one of those where the bridge doesn't collapse: unfortunately we cannot (yet) move from one section of the multiverse to another... However, the split between classical physics and the probabilistic/aleatory subatomic behavior also seems to reproduce itself in the way we design/transform our environment, but with the further problem that there is no longer a matter of scale and these approaches tendto collide. The continuous that crosses our contemporary societies is fully embedded in our practice and in our thinking.

Hard sciences are intrisically counter-intuitive (our senses tell us that it is the sun that moves around the earth). On the one hand they taught us to set up paradoxical strategies, using chaos to produce open and dynamic orders, looking at self-organization as a possible and more effective (desirable) horizon. On the other hand, we do not have to prove our hypothesis: we just tell stories, and science is a big reservoir from which to fish for powerful devices, tools, and arguments, in order to infiltrate opinion, building the conditions to make our strategies really work...

Benoît

There could be no more big stories, and small stories could not be considered as serious or efficient palliative treatments.[13] (And by the way are we really sick? And does it matter? Isn't sickness a form of living? Not to be associated necessarily with death, but as an active agent of life and transformation. Death didn't exist on Earth for a while; it appeared as one of the possible developments of life on Earth.)[14] No strategies are ethically good enough to be immunized against distortion. Deleuzo-Guattarian theories could be used either by architects or by military strategists submarines, while the other is the Gulf Stream, the most energetic mobile trash can from Florida and Europe to Scandinavia. Confronted with these extreme conditions, including these new parameters, the polar bear transforms physiologically, a mutation we see in real-time as a Darwinian effect in the wild. Their hermaphrodism statistically increases their chances of reproduction and survival. The 5 percent of mutant post-polar bears are Houellebecq's characters—brother and sister, parent and children, female and male—modifying their comportment, their sexuality, to adapt and recontractualize their link to the environment. They are not denying or theatricizing global mutation, they are absorbing and integrating the mutation as a new protocol, as a new contract, as a Sacher-Masoch deal.

Joker

Well, well. Are you sure you want to introduce Sacher-Masoch as a value? Do you want to contractualize with the devil, through his Mephisto servant?

Benoît

You can do as much contractualizing as you want, there is no limit, no special restrictions relative to that, but the durations have to be defined. From the peak of the Ras Dashen (playground of >

boundaries of knowledge,or better, the boundaries of interpretation of knowledge. One way with Björk and Disney entertainment, naive and illusionist, and the other way with Golem, Gollum, Mephisto, and other middle-aged characters, freak baby dolls, and self-negotiations with the unknown created by profane technologies... as a de-sacralization of the science and its application. Why choose between Yoda and Darth Vader? They are just Siamese-twin sisters from the same matrix.

f. r.

the gods) to the Schwartzwald (another one), from the dark dancing of Karachi to the gas-fuelled isbas of Irkutsk (different kinds of human heat), there is an infinity of thresholds, entrances, gaps, lost corridors, and hidden passages where this kind of ceremony could unfold, no fear of slicing my thumb with a razor to ratify the pact. But first I would like to choose the territory, a topology that we could all agree on, and a defined area where it could occur. Not an especially comfortable one, not well known or heartening—we all know that it has little to do with pain or satisfaction, all resides in where the contract will be made, not the terms of it: all you will remember is the place where you made it, the taste of snowflakes on your tongue, or the sweat on your flesh at its contact with fur. But there is no contractualization in the usual terms; it should not be confused one shot of time, with a rewind and fast-forward story. The uncertainty of the system is something we definitively approach, even by crossing to the dark side, by revealing the ambiguities of a situation. Look at the opposition between sponge geometry and Hippodamos's master plan of Miletus. This opposition is clear: an open system, where the algorithm of growing cannot be reduced by a simplified relationship, and a closed system, coming from the architecture field, as the first planning, where everything is predictable, forecasted, and frozen. The opposition is not only a difference of processing, between algorithmic and Euclidian. The sponge another, trying to find a path through modelization or programming. If an unexpected or "emergent" event arises, the function that you conceptualized a priori is totally disrupted (and I'm not even talking about trying to input two or three relationships of cause-effect at the same time, only a single one, or even when calculation leads to undecidability or several results). Introducing fictional material into a process is a way to spread "intelligence" to the whole movement and to be able to react at each step, to evolve with the project. Biologists have been searching for decades for the pacemaker in slime molds, thinking that they didn't have enough information to find it, that its disguise was keeping it unknown, to discover finally that there was no pacemaker, that all the cells were able to decide when to unite and when to separate.[17] The consideration of panoptical or "transcendental" supervision sounds always a little bit vain to me in architecture or in urban planning, just like an illusion of control, of cause-effect relationships in length and duration.

Giovanni
In school, we were proudly told that Brunelleschi became the first modern architect when he fired the workers of his big dome in Florence. Since he was the only one who knew how to build it, the shared knowledge of the medieval building process was not working anymore, and he could rehire the same men for less money, as an underskilled workforce. So, modern architecture was born with that act of domination, which followed the increasing complexity of social, economic, and technical processes, where complexity meant separation on multiple levels (and maybe someone remembers Marx and his theory of alienation). What is interesting today is that there is such an S&M, multifaceted relationship between architects and the other social subjects involved in urban negotiation (developers, politicians, builders, users, citizens). It has become unbelievably complex and fragmented, in a way where vertical control is no longer a viable approach. François is right

apparatus 3: physiological mutation

Polar bears in the north pole are mutating into hermaphrodites because of pollution—an acceleration of Darwinism from global warming effects—a direct negotiation in wildlife with this danger. The polar bears don't deny the evolution of the climate or emphasize it to proudly appear as a "Don Quixote" saving the world. The polar bears negotiate and absorb the movement as a physiological and sexual mutation to readapt themselves. Could we renegotiate our behavior, neither in a moral way, to clean the planet, nor in a catastrophic way, to wash away our recognition of the drama? No coward in the sympathetic denunciation, no bastard in the cynical investment to profit from the situation…

f. r.

with Faust or Sade. There is no possible liberation or emancipation, don't even think about it. It's a necessity and an irresistible urge; it requires disguise on both sides; a biotope and a disguise.

François
The Sacher-Masoch apparatus is previously defined by protocols; it contractualizes and defines relationships, which become the frame, the rules of the game directly dependent on the nature of a contract. But at the same time, as a transitive process, the writing of the contract defines as a preamble the condition of the instruction, introducing the boundaries between acceptance and erasure, between what is legitimate and what you have to renegotiate in case desires drift during the game itself. It seems more contemporary than the deal of Faust, where the contract is definitively the agreement of your dependencies as a plan of no return. We are with Sacher-Masoch in a reflexive alienation, which introduces the re-enunciation of the condition of emancipation at each step of its own evolution.

Joker
Please come on, where is the architecture field in these "Sturm und Drang" speeches? Do you think you are focusing on something or just digressing from nowhere to nowhere?

François
Well, well, you shot me, nasty Joker. But the endlessnessless is coming from this kind of apparatus, with an adaptable and readaptable, open-source system, depending intrinsically on the mutation of the system. The main question is: How could we develop open protocols, able to include a wide latitude of freedom? Or, more precisely, where could the system develop its own generative evolution to absorb and react according to the mutation of preliminary parameters? From the house in the forest, a long-ago project, where the growing of the trees slowly destroyed or weakened the house, to the robotic apparatus of I've Heard About, we always considered the scenario as an open narration where the architecture should be just one element, does not depend on a stochastic equation, but on a kind of cognitive process of induction deduction, adapting its own growth to its global equilibrium (static, photosynthetic). In this case the sponge is not dealing with mister Sacher-Masoch or Mephisto, but it integrates the unknowledge of its achieved shape as a value of its own existence. It radically and politically changes the condition of emission: it could modify the status of the architect, who, if he runs in this profile, will become an alien progeny from the incest between Villard de Honnecourt and Filippo Brunelleschi.

Benoît
It cannot only be understood as a game of "possibles," even if it has a lot to do with probabilities. We know that even if we can prove the probability that an event could occur, and by the way determine the relation between A and B, we will never be sure that it is the right and only one. To be more pragmatic, let's take two different states of a shape in a numeric process. Let's say that we have a topology and a function, and that we want to find the way to pass from one to when he underscores that we are now playing, as architects, both the characters of Villard and Filippo. The problem is where and when self-organization and control occur. Indeterminate devices, diagrammatic routines, open-ended scripts, and so on are often strategic means to define a set of conditions where we can still be architects, where our specific knowledge does still make sense.

Joker
You look nostalgically to both Brunelleschi and Villard. But it is evident that there are no sacrifices that we can make to be absolved of the original sin of modern architecture, and the knowledge that made us what we are can unfold completely only in fascist situations: Dubai and China are now the architects' paradises on earth…

Giovanni
I'm just saying that our condition is to negotiate architectural choices within >

apparatus 4: ecological anomaly

From Kurosawa's movie *Charisma* comes the debate and fight between ecological ideologies, between a dinosaur tree—weak but highly toxic, a survivor of wild nature from the past, sending toxins into the ground in the dying biotope around itself—and, at the opposite end, the domesticated nature of a cultivated and economically profitable forest—a domesticated nature that the weak dinosaur tree is killing. The little creature finds a way to protest its eradication by fire by exploding into a nuclear green mushroom…

f. r.

indeterminate environments, and vice versa. Look, for instance, to the opposition between the sponge and the grid, where the first is the outcome of a self-organized process and the second a simple act of top-down planning. Are we sure that a spongelike urban structure is more indeterminate than an orthogonal one? It is not just a matter of representation (organic vs. geometric), or an issue tied to the design tools we use. A

with the CIA for a torture scene) and another from *2001: A Space Odyssey* (designed with NASA for a Galilean escape).[20] It seems difficult to simplify this reality, to reduce it to a simple, univocal game, without considering the ensemble of heterogenic tools integrating speeches, regulations, strategies, scientific protocols, games of power, and stories of self-alienation; integrating talks, nontalks, and misunderstandings

apparatus 5: planning and self-organization

We can see, in the opposition between these two pictures below, two systems of organization: from Hippodamos's master plan of Miletus to the unpredictable growth of the sponge. Both are driven by algorithms and genetic orders, but one is fully forecastable and closed in on itself, and the other is open source, readaptable, reprogrammable, depending on the mutation of the environment and all the constitutive inputs of its own mutation. The first comes from the criminal pride to master plan and use its animation as a system of control, of individual and collective survey, in the sense of Michel Foucault. The other paradigm accepts taking a risk, a political risk, an aesthetic risk, to "open source" in a going-to-be-done movement, as an endlessnessless...

f. r.

grid (generic) can work as the framework for very indeterminate behaviors, while a sponge (articulated) can trigger very specific local answers. I think that we should take a fractal point of view, with alternating layers (natural/artificial; Euclidean/non-Euclidean; controlled/self-organized) that depend on time, scale, 2-D/3-D shifts...

In other words, to go beyond Villard and Filippo, we have to merge them. If

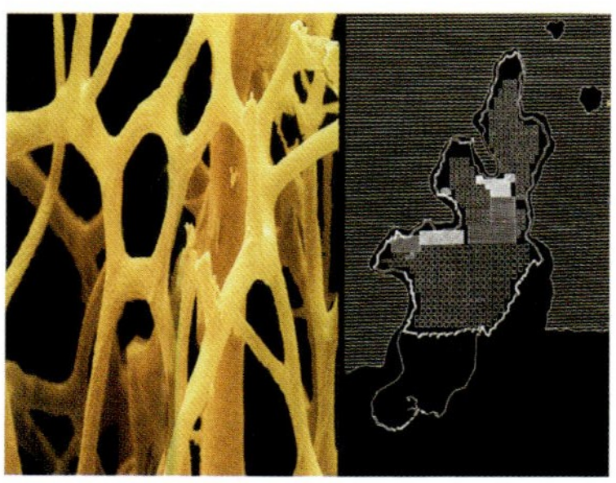

contemporary attraction to the future attached to the 1960s through formalist derivation, the reduction of the future exhibited can be considered simply as an aesthetic decision reflecting the taste of the author, its uniqueness. But from futuristic content translation to the intensity of revisited fiction, or the blooming dystopias, such fascination seems to flirt only with commercial business. Is there no exit? No, the future is gone, as Ballard claims, as well as the technological potential embodiment attached to it.[23] I don't think there is any legitimate space left for refuge.

Benoît
Yes or no, the future is gone and will never be. Climatologists are the first ones to know that they will never be able to make accurate predictions even at mid-term, even if they will one day own computers as fast as the demon of Laplace.[24] Actually, they prepare themselves for the uncertainty of climatic changes by elaborating on families of scenario. These scenarios deal with hierarchies, but more interestingly with heterarchies, of information.[25]

François
Our digression or chat, turning around our subject, works as a very careful approach, without cadavers in the closet. The speech seems to be on a very strategic point. As we said, endlessnessless is a tool for narration, for uniqueness, not for emphasis of industrialization or repetition. It includes and produces scenarios of singularity, of anomaly. Endlessnessless redefines the aura of items. In a way this narrative machine—extracted physically, screw by screw, from mass industrialization—is able, first, to escape from this genesis, and secondly, to develop both stories and principles of reality, application scripts, constructive behaviors, and impermanency, uncertainties. This kind of post-bachelor machine, coming from the incest between the T-1000 and Francis Picabia's ghost, introduced subjectivities in the touchable substances of physical transformation.[26] This is one of the ambiguities, but not the

the first was inside the process and the second designed the object, we now have to design the process.

François

Many indices are whispering to us, from *Architecture Without Architects*, written by Bernard Rudofsky in 1964, to Frédéric Migayrou's analysis of the "dispute" between van de Velde and Muthesius in 1914 as being between industrialization, series, and prototyping as the identification of the uniqueness of a situation.[18,19] This debate has re-actualized since the 1980s, because of the two bad "golden boys" Steve and Bill, who democratized the tools of control and narration.

We could consider this genesis period a frozen one, dreaming of the last retro-future building as a Zaha-homage-vintage-positive-white-future item, pre-designed in the 1960s, but strategically constructed half a century after.

I know it's painful to recognize that the future drifted in another direction, that it's a lost sensation. The current period is a time sandwiched between a predictable future that never happened and the unknown of tomorrow, which is coming day after day, something between *In the Mood for Love* and *2046*, both by Wong Kar Wai. This sensation of erotic dystopia, of charming distress, a Baudelairian spleen according to Walter Benjamin, is a perfect reversal of the blooming of modernity, where the lost paradise emerged from the indistinction between mass production and the production of the mass, where the loss of uniqueness opened the overvaluation of repetitions and series, disqualifying anomalies or singularities as illnesses of the system. At the opposite end, the spleen of today does not come from loss of value, but from the impossibility of attributing value to this uniqueness, definitively lost after the death of the experiment of modernity.

How, in this condition, can we hide ourselves, somewhere in a comfortable back room? The conditions of today, here and now, are palpitating between dream time and the day after, as altered states mix with schizoid ingredients, with a pinch of *A Clockwork Orange* (designed as the network, the rhizome of the narration, of the scenario, as preliminary enunciation of the apparatus of an architectural item, and, where the apparatus cannot itself be reduced to the definition of the architectural part, where the input and output became contingents, where ambiguities are articulated between themselves, where the protocols of transformation reveal the condition of emission and are revealed themselves as Situationist strategies.[21]

Alexandra

I'd like to go back to the famous 1914 acrimonious polemics between Muthesius, who was previously a spy for the German government while working in England at the turn of the century, and van de Velde. Though far away from the temptation to give a romantic analysis of the standard ("good" at the beginning and then struggling to gain importance as the First World War started, as Nikolaus Pevsner claimed), the cofounder of the Deutscher Werkbund was more of an idealist than van de Velde. Individuality vs. *typisierung*? Pros vs. cons? Authorship vs. standardization...? I don't think so! Regarding the economical shift, the latter was more pragmatic—and this explains why Gropius, Taut, and many others followed van de Velde in his ambiguous contradiction—but it is the former who obviously took historical advantage. Besides, the most important fact of the debate consists, according to Frederic Schwartz's assertion, in the emergence of the copyright notion, and of the artist, architect, and designer as the legal equivalent of the industrial: "Muthesius invokes a central point of copyright law: the right of the author to have his name appear on or next to his work (or, conversely, to withhold it), even when the work is executed and sold by another party. This was the true polemical gesture of the Dresden exhibition and the source of the controversy which led to the founding of the Werkbund: the central symbol of copyright."[22] This controversy indicates a primary discussion about mass culture, and mass-mediatic culture, as it is published and exhibited toward an audience.

While identifying the futurist- main one: at the middle of these huge possibilities of prototyping application, we are paradoxically fully infiltrated by a spleen, confronted with the difficulties of whether to apply or define value to this uniqueness we could realize. It works as the lost sensation described by Baudelaire, but in a perfect symmetrical direction; as kids of *Tron* (computer nerds swallowed by the software), we are able to realize a prototype, rare and unique, with technology, but the genesis of this prototyping product has been digested by a Miyazaki monster and lost by the system of growth of desire.[27,28,29] It's a very strange state that, while we could be involved with computational design, in a nonstandard approach, the intrinsic senses and values of this approach and production drift and shift somewhere, in a magnetic black hole.

We can suppose that the emission of the uniqueness cannot work as the reflexive alibi of its existence, as propaganda of its loneliness...

Joker

Its seems to be pre-romanticism talking—like the Sturm und Drang Goethian movement—against the rationalism of the Enlightenment and the positive aspect of the encyclopedia. But architecture has always been projected with these kinds of positive aims and purposes, as a vector of progressive projection. Are you at the opposite end, regressive and pathologically alienated by the Faustian deal?

François

Do you know the sad story about the mouse dancing provocatively in front of the cat, only to be killed and eaten by the domestic feline? Apparently, the condition of provocation is fully nonunderstandable and illogical. It seems to be a suicide, with music and elegance. This could be an illusion of a pact with the devil. The spirit of the mouse will survive on the condition that it be digested by the cat. But in another way, this is the story about little bacteria, the Toxoplasma Gondii, which need to live in the stomach of a cat to reproduce.

This strategic microbe first infiltrates the food of the rat, and moves from >

the intestines to a region of the brain, to disinhibit the rat's fear of the cat, forcing the victim to dance in front of the predator. Absorbed by the cat, it finally finds the ideal biotope of its reproduction.

We are talking about something that appears absurd, in a rational way, in an allusion to the dance of death, but this story stays at a level less monstrous than the causality and dependency explanations of neurobiology science, as a loss of free will, of independent consciousness.

Benoît
Those undercover and nonlinear trajectories are completely changing the approach that we have and will have to all fields of human applications. Genetics are not based on a regular timeline, but on an imbrication of causalities and a/effects (affects and effects as one single thing), all reversible, possibly postponed, and fragmented. The popularity of the ideas of a simplified Darwinian evolution contaminated minds almost better than this bacteria.[30] Recent research in biology pushes one to think that there is no "intelligence" other than a silly one.[31] If you put bees and flies in a bottle stuck on the ground, the bees will die after a few hours, exhausted from bumping on the glass, their instincts telling them to go where they see light; conversely, flies will find their way out sooner or later through the neck of the bottle: their apparently erratic, uncoordinated flying being much more efficient. But there is much to say about the a/effects. If we can determine the cause of relationships by studying them, this comprehension leads to no real knowledge and understanding.

Knowledge and its application, if systemized, could not be replicated. Pierre Gourou, a French geographer, went to what is now Vietnam in 1927 to understand why the population density was so high.[32] Looking at countries situated at the same tropical latitude and at the kind of agricultural exploitation the weather conditions and topologies were inducing (a nonintensive one), this density was totally anomalous. The population of the country should have been much lower. He discovered world like ours. We should talk rather about variations and singularities. Monstrosities as extreme singularities redraw the boundaries of the intelligible world, redefine the norm, and rescale our tools. Georges Canguilhem was one of the first to think that searching for them in the history of science permitted one to draw much better maps of the thoughts of a time than its official theories.[34] What is fascinating is that at exactly the same time of the argument of Muthesius and van de Velde, a strangely similar polemic occurred between two scientists, Max Planck and Ernest Mach. Planck accused Mach, through his historical-pragmatic vision of physics, of threatening faith in the intelligible unity of the world. Both Mach and van de Velde were trying to deal with those singularities. Gropius, probably still under the influence of Muthesius, drew the first logo of the Bauhaus as a cathedral where all artists, scientists, and thinkers could reunite themselves. Quite eloquent, no? Still and already the belief in a big secret order against small disorders. It leads us to our frozen time. *Aufklärung* vs. *Weltanschauung*.

Giovanni
I think that we can agree on the fact that to project means to deal with cause-effect relationships. It may sound like an enlightenment or positivist statement, but it is a condition that is very hard to escape. On the other hand, we know that deterministic actions are less and less likely to get the effects they expected. (There is a joke about TV ads in Italy: some years ago, a liquor firm promoted its product with the adventures of a veterinarian. They didn't sell one more bottle, but the veterinary schools had a lot of new students.)

The majority of architects tried and try to resist this situation, since they feel all this indeterminacy to be a deadly threat to the core business of the discipline, i.e., authorship and formal control over the built outcomes. But reality escaped them and the occasions where the architect can actually rule over the building process are falling down (one of the last issues of *Volume* shows

One of the weaknesses of the "nonstandard"approach (especially by the Americans) resides in its lack of direction or, in other words, in its search of novelty as an absolute value.[36] It seems that many outcomes are simple consequences of technological possibilities: answers in search of a question. They act like gamma rays (or the flies in the bottle), producing a large amount of alternative solutions waiting for something (critics, magazines, the market, clients?) to select the next architectural "real thing." It is not so different from what happens in the turbocapitalistic development of the eastern tigers: I've heard that in Bangkok they thought to build two different metro lines to serve the same area, waiting to see which one will survive...

Benoît
We are, for sure, now in the area of cognitive capitalism, all the fields of what is "free" time or "free" production are reabsorbed and instrumentalized: to justify positions, roles, and to nourish contingencies. But to refuse to empower any direct cause-effect relationship for a while—to consider them as something purely illusory, only made to feed hollow simulacra and replicas—helps us to avoid such instrumentalizations. In the first books of her series about the history of science, *Cosmopolitiques*, Isabelle Stengers talks about a scientist of the end of the eighteenth century, Lagrange, who wanted to establish a model that could determine every state a priori.[37] The rules of this model were established through empirical experiments. All events found during experiments that could not be integrated into the equation were pushed aside. There was no more causality in itself, only states and equivalence of forces between those states. Just as if an object was frozen in a certain state of friction, balance, etc. Everything that was able to attain this state and everything that would come after was calculable, in a very deterministic approach. His main problem was that each singularity that interfered during the experiment was a threat for the whole system. About the Lagrangian system Truesdell said quite ironically: "Nature is begged to adjust

and proved that it was the "culture" and the society of the people that allowed the population to grow in this unexpected manner. It was not due to a system, nor to a special technology, and most importantly, it could not be reproduced, even in similar countries at the same tropical latitude. What we today call externalities created the conditions for this unpredictable human growth. Because of the difficulty in calculating them and knowing their precise role, externalities are hardly integrated in soft or hard practices. But what we call noise (or small music: the ventriloquist's song of the bacteria in the mouse) is completely transforming our understanding of physics and social phenomena, for example.

Even if a system could not be replicated, it does not mean that uniqueness could exist.[33] Something unique could not be a part of a factual >

that even celebrity doesn't assure the final cut).[35] So, in the last twenty-five years architectural avant-gardes moved their struggle against the establishment from the language field to an operative one, dealing with indeterminate process, bottom-up techniques, open-ended devices, and interactive protocols. They triggered a conceptual shift that brought proliferation—rather than composition—to the foreground. But proliferation alone is not so clever. It is an accelerating tool, like the gamma rays used to mutate wheat seeds. They have to sow them and breed some generations to understand which are the fittest. More interesting, then, are the parameters of selection, the conditions with which to negotiate. That's why we all love the nonpedigreed architecture, for its sense of necessity, and that's why many of us hate it in its recent market-driven, global, generic, repetitive expressions. >

itself to the mathematics rules learned on school's benches."[38] It was probably considered by Lagrange himself a default that could be solved by a matter of time. Conversely, at R&Sie(n) we take those singularities as the main subject in our architectural practice, as the most fertile ground. No longer a pure arrow of cause-effect, but a succession of frozen objects, in different states, loaded with all the "affects" that it went through, made of all its footprints. This manner has the advantage of not letting us sink into a world only ruled by probabilities. There is this fascinating book by Philip K. Dick, *Solar Lottery*, whose story takes place in a world entirely controlled by probabilities: "People lost faith in the belief that they could control their environment; all that remained was probable sequence: good odds in a universe of random chance."[39] The most interesting part of this world was how human minds were able to create the machines that were calculating those probabilities, and how human minds could face the uncertainty produced by those machines (and cheat with it, of course, we are in the paranoid world of Philip K. Dick). Those two Bangkokian metros make me think about the Turing nondeterminist machine, which allows several results for the same calculation: negative, positive, or indeterminate, all of them being "non-untrue."[40]

François
I remember the voice of Antonin Artaud in the borderline radio play *Pour en finir avec le jugement de Dieu*, registered just after the Second World War and forbidden for forty years.[41] Screaming, whispering, crying, he was vomiting about how "les americains fabriquent des machines et sont fabriqués par celles-ci." I deeply understand this machinist claim at a period when Cybernetics and AI (artificial intelligence) were developed as ideology. To the contrary, Turing, the British mathematician who cracked the Enigma (the German Nazi code machine) with the first computer he developed, killed himself in a similar way to Snow White in the fairy tale, by eating an apple after having bathed it in a cyanide solution. The bachelor machines or apparatus >

apparatus 6: the day after and dream time

Two representations of time: the NASA one, with a dream-time effect in the *2001* odyssey for Promethean research, and the CIA one, with a day-after sensation in *A Clockwork Orange* for torture knowledge. The notion of time is a blur, and its arrow hesitates between these two directions. We cannot deny that the future is a feeling coming from the past, vintage and nostalgic. In this way, the future cannot be the future, but something else, with another denomination, or with no denomination...

f. r.

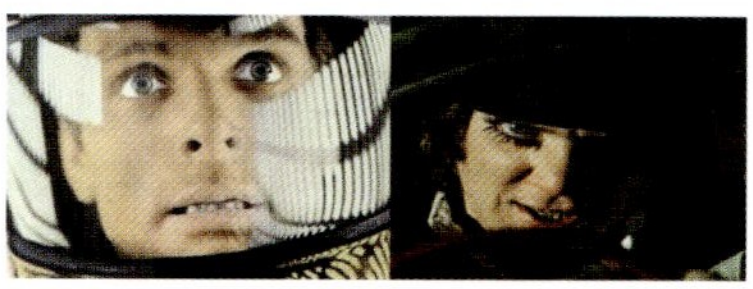

articulated by R&Sie(n) borrow more from the possibility of machinism as a protocol of subjectivities, and borrow some tools from scientific, technical, and technological backgrounds—from which emerges a kind of indeterminacy that instrumentalizes the failure of their positive nature and their structural logic. Ghosts are always hidden behind and within and beyond scientific practices, intrinsically.

For example, when Marcel Duchamp discovered Xenakis's music and Philips Pavilion (and, accessorily, Le Corbusier), he hypothesized that these productions were infiltrated and contaminated equally by the influence of mathematics and little girls.[42] Beyond what seems to be a provocation, which brings us back to the starting point of this chat, he was suspecting the illusion or mystification of the theory of correspondence (from Varèse's sounds to the definition of space in the UPIC software of Xenakis).[43] If we take the risk and go further, Deleuze and Guattari passionately developed this strategy of subjectivities as a strategy of resistance.[44] The scenario of subjectivities and the constitution of singularities are the two keys from these twin brothers that helped them to re-scenarize a situation and escape from the exclusive determinism of pure computational addiction. Strategies of resistance could be reread from the Maoism movement of Robert Linhart in the seventies, called "Les Etablis."[45] When the group decided to stop teaching at Nanterre to infiltrate a capitalist car company, the purpose was to reinforce from inside the social class struggle by organizing a worker strike, to dealienate the conditions of the product produced by the mass. But they never tried to reconsider the design (of the car) as the vector between the mass production and the mass desire. The apparatus of resistance they introduced was only linked to social conditions or political value. They never tried to articulate the design of the car as contingent to the situation of mass alienation. They never accepted the whole situation with all ingredients of the system itself, from touchable parameters to psychoanalytic reinterpretation of

fiction to enter, to create the openings, to be an instrument of transformation. Pure computational determinisms preserve their authors from this risk, allowing them to refuse speculation because of the impossibility to have full control over it. But speculation is a dynamic and physical object, it's always in motion, it's subject to the Doppler effect: it's by its taste and color that you can recognize and understand its shifting.

depicted in Roussel's novels finds a parallel organization in Duchamp's *The Bride Stripped Bare by her Bachelors, Even* masterpiece. Male at the bottom and female at the top: no reunion, no union—they can't mate. Hugo Gernsback, the father of modern science-fiction, could not resist dedicating an issue, its cover, and several texts of his magazine *Amazing Stories* to the scientific mating of human beings through machines.[49]

apparatus 7: causality and dependencies

The Toxoplasma Gondii are bacteria that need to be eaten by a rat to remove its fear of cats from its brain cortex. The rat is forced to dance in front of the cat and be eaten, and at last to travel to the stomach of the cat. This is the only place where the reproduction of the Toxoplasma Gondii is possible. We are just running around something that appears absurd, and we have to resist the temptation to use science in a positive way, to hide the reality of this nonsense, driven by causality and dependencies that never appear in the visible spectrum.

f. r.

Alexandra
I am back with the bachelor machines. The sexual frustration attached to the machines provides a line of thought that starts with Villiers de l'Isle-Adam's *L'Ève future* and finds, with the dystopia zone of R&Sie(n)'s machines, a new stance.[47] Let's go back to one of the sources with Raymond Roussel's *Locus Solus* or *Impressions d'Afrique* texts.[48] The latter was performed at the Théâtre Antoine in Paris in 1912, at the sole expense of the rich and eccentric French writer. The

Bachelor machines don't give a damn for cloning or duplication…they crave pleasure, only pleasure! In addition, I would like to bring the perspective and the filter of Samuel Butler's novel *Erewhon* (1872).[50] Butler depicts the

these parameters.

Benoît
I've caught my breath. I was bewitched by the Joker for a while (that sort of thing happens sometimes); I want to talk about real time, which could sometimes be related to augmented reality. Some of our friends, by refusal of any retro or techno future, want to live in illusory, perfect suitability to the present time. It leads them to believe in the precision, the efficiency, and the honesty of the tools that they have access to.

Even if we can't consider it as plain positivism, it leads to a sort of scientific mysticism, understandable by what it tries to avoid, but not by what it generates. Ballard declared in 1962 that "The only truly alien planet is Earth."[46] He had been well known for years as a heroic sci-fi writer, but in later years chose to get back to writing about human pathologies.

The literary approach known as "speculative fiction" never works on a perfect timeline, there are always subtle shifts, small gaps in reality that allow >

whole avant-garde, from Duchamp to Breton, attended the show. And, as they were stunned by the various artifacts, it is no surprise that the sexual nature of technology became such a paradigm in their respective works. It constructs an interpretation of a machine symbolizing pure bliss through both onanism and a sexually liberating and nonrepressive, sadomasochist pact. The whole range of the bachelor machines borrow from the coitus-interruptus process. It has, at its heart, the idea of sole pleasure; it prevents the deposit of sperm into the vagina and therefore focuses on bliss and rejects production. The energy and tension of a bachelor's counterproductivity as >

mating system of machines: "Does anyone say that the red clover has no reproductive system because the humble bee must aid and abet it before it can reproduce? No one." Machines use humans to "aid and abet" them. As John Langrish claims, machines enslave individuals to ensure quality.[51] Behind this bizarre reception of progress and evolution is more than a reconciliation: it foreshadows and considers a loving, sadomasochistic relationship between men and machines. We vacillate between the impetus for technologized power and the fear of machinery versus individuality, as expressed in the seminal play *R.U.R.* or in Asimov's Three Laws of Robotics.[52] What comes next? A blind spot of vulnerability and pleasure again! Such a legacy is necessary to fulfill R&Sie(n)'s machines. Orgasm is achieved through the power of the mind (the bride and the bachelors of Duchamp's *Large Glass* accomplish and culminate their mating mentally) or subconsciously: "The subconscious is a factory, a machine for production," as Deleuze puts it. A genealogy of sluggish and desiring machines and machinations, from Duchamp to R&Sie(n), arises.

François
In the scenario of R&Sie(n), the "bachelor apparatus" has to be assimilated in a vector of narration, like the lyre of Orpheus when he went down to the kingdom of Hades to bring back Eurydice, his sweetheart, and played music to bewitch the wild animals and the devil. The apparatus creates an operative effect and a blur logic at the same time. What kind of music could be played now? Stockhausen or Sirtaki, John Cage or "Just Like a Woman"? What enchanting, magical effect could come from the instrument to pacify evil?

Simultaneously, the apparatus build ambivalences, both narratively and procedurally, as schizoid contingencies, and the machine becomes the vector of a constructive subjectivization.

To tell a story about architecture, R&Sie(n) have introduced gimmicks and MacGuffins.[53] For example, the mathematical formula in the movie *The 39 Steps* is a clue from which the story >

apparatus 8: unpredictability and chocolate

Kinder Surprise candies were forbidden to be sold in Los Angeles in the sixties because of the toy inside, which was considered an item in contradiction with American values: you could buy what you desired with credit, but with the condition of knowing exactly what you were buying. These unpredictable toys, hidden in the chocolate eggs, were denounced as bad influences for kids, but in the same period, LSD was freely accessible...

f. r.

apparatus 9: bachelor machine

Duchamp's *The Large Glass* is a machine of loneliness, the opposite of a mass-production device for standardized production. Machines here are a tool for narration, not for emphasis of industrialization. They include scenarios of singularity, anomaly, rarity, and, paradoxically, a sensation of the spleen, in the sense of Baudelaire, of course. The machines could redefine the "aura" of Walter Benjamin, by two narrative woven parameters: the uniqueness and the stories embedded in the production process itself. The machines are vectors of transformation. They are application tools but also vectors of rumors, where the impermanency and uncertainties of their real function and behavior introduce subjectivities into the principles and protocols of realities.

f. r.

v

could have unfolded, but the film releases itself from this origin and becomes, after a while, independent of this clue. The Olzweg machine for the FRAC is the starting point of indeterminacy, of aleatory behavior and a process of losing control over an unpredictable shape. The endlessness needs this clue, both narratively and operatively, to create the condition of a further step, of an unachievement, of an after death experience, coming as a prolongation of the phase of the construction. For this, R&Sie(n)'s apparatus use stochastic machines, psycho machines, chimera robots, speculative mechanics, anthroposophic systems, depolluting

processes, environmentalist ecosophic devices, and paranoid artificial climates for negotiating with biotopes' fears. The machine protocols are psycho-masomachinistic, including misunderstandings (*des malentendus*) and frustrations. The level of freedom, the latitude of randomization of behavior, is developed as corruption of the application (for instance, the use of inhabitants' stress as a transformative parameter in I've Heard About). The script, the algorithm that drives the machine, is disrupted by internal agents dealing with "if, then, and while" possibilities and alternatives. But, as I mentioned, the main aspect of the apparatus is to reveal, to release the contradictions of a situation. They are not simplifying and denying the preliminary complexities, but they define a strategy of mutation directly extracted from their recognition. Are they desirable and desiring, in the sense of the body without organs, defined by Artaud and Guattari?[54] I hope so. Their eroticization, their sexualization—weeping from the context where they are embedded—are a part of the rhizome, to avoid a panoptical unfolding. As Pessoa wrote in 1952, "I'm coming from before the reality."[55] The apparatus placed the body of architecture in a space that combines the real and reality, using fiction as the traveling vehicle.

Joker
Are you playing music, like the Titanic's endless song?

François
The nature of the music played during the sinking is still under debate: some survivors heard "Nearer, My God, To Thee," while others heard "Alexander's Ragtime Band." Why did witnesses' memories alternate between the ideal request for a song of God and the request for profane ragtime, between the deal of death and the last human rage against the deal? So, if you want to compare, dear Joker, R&Sie(n) is improvising both simultaneously.

But don't be confused, the lyre is a decoy, a lure. It assumes the operative and visible part of the apparatus is cooked using a more sophisticated recipe, with a pinch of ecosophic empathy—as the recognition of a preliminary situation—a pinch of a Sacher-Masoch contract—as a rule of the game or scenario—a pinch of an anthroposophic loop—as an exchange of substances—a pinch of a heterotopian sensation—as a protocol of indeterminacy and stochastic behavior—and the last one, the cherry on the cake, a pinch of a dynamic agent or lyre—to operate the story, from Orpheus's hands.[56,57,58,59]

Joker
Yes, yes, you've got it! The five points?[60] Again? As the nighttime revival of the zombie?

François
You killed me, my friend. "I'm late, I'm late, I'm late!" said the White Rabbit. It's time for Alice to jump into her parallel universe. Like her, now, we may confuse our own paranoia with the unreality of our perception.

1. *The Tempest*, ed. Robert Woodrow Longbaum (New York: Signet Classics, 1998), 1.2.195–200. References to act, scene, and line.

2. See Giorgio Agamben, *Che cos'è un dispositivo* (Rome: Nottetempo, 2006), 3. "The hypothesis I'm going to propose to you is that the word 'apparatus' was a technical term in Foucault's thought strategy."

3. There is a genealogy of rabbits that remain to be mapped out from Fibonacci to Lewis Carroll. Its title would be *Rabbits never die...*

4. Heraclitus frag. 93 Diels-Kranz. "The lord whose oracle is at Delphi neither speaks nor conceals, but gives signs."

5. Attributed to Thales of Miletus.

6. See Georges Bataille in Denis Hollier, *Against Architecture: The Writings of Georges Bataille* (Cambridge, MA: MIT Press, 1989).

7. René Thom, *Morphogenèse et imaginaire*, Circé: Série Méthodologie de l'imaginaire (Paris: Lettres Modernes, 1978). "The human nervous system is an organ of alienation. It can be something other than itself." See how the embryonic blastula invaginates itself producing the tri-blastic structure of the 3-derms.

8. Henri Bergson, "Le Souvenir du present et la fausse reconnaissance," *Revue philosophique* (December 1908): 561–93. English translation available in *Mind-Energy: Lectures & Essays by Henri Bergson*, trans. H. Wildon Carr (London: Macmillan, 1920).

9. Baruch Spinoza, *Ethica ordine geometrico demonstrata* (1677), part III, proposition 54.

10. The first time that Deng Xiaoping used the "cat theory" in public was in 1962, at a meeting of the Chinese Central Committee Secretariat, when discussing the "contract responsibility system" for restoring agricultural production. Curiously, at that time, the first cat was yellow...

11. Gustave Flaubert began *Bouvard et Pécuchet* in 1872. It was published unfinished in 1881.

12. Brian Aldiss, *Cryptozoic* (London: Faber, 1967).

psyche and the soma was unfounded and untrue. On this topic see Antonio R. Damasio, *Descartes' Error: Emotion, Reason, and the Human Brain* (New York: G. P. Putnam, 1994).

26. The work of Picabia stays in the shadows of the twentieth century, but his kaleidoscopic ghost is always perturbating the hierarchy of values. Here we refer to the exhibition in 1922, Máquinas y Españolas, at the Galeries Dalmau in Barcelona.

27. Charles Baudelaire, "Loss of a Halo," in *Paris Spleen*, trans. Louise Varése (New York: New Directions, 1970).

28. *Tron* is a 1982 film directed by Steven Lisberger.

29. See the black digestive intestine in Hayao Miyazaki's 2002 film *Sen to Chihiro no Kamikakushi* (*Spirited Away*).

30. In a very chauvinist way—still and again the Fachoda complex—the ideas of Darwin on evolution were influenced, mainly in France, by the theory of Lamarck, a French scientist of the early nineteenth century. Lamarck's theory was that acquired characteristics were created by hereditary transmission, while Darwin's theory proposes the conflictual relationship between the logic of organisms and their environments.

31. Mathieu Aury, "Darwin révolutionnaire? Une lecture politique de Dennett," *Multitudes* 16 (2004). English readers could see Daniel C. Dennet, *Darwin's Dangerous Idea: Evolution and the Meanings of Life* (New York: Simon & Schuster, 1995).

32. Pierre Gourou, *Les paysans du delta tonkinois* (Paris: Les Éditions d'art et d'histoire, 1936). Also see Gourou's contributions to *Dictionnaire de la géographie*, ed. Jacques Lévy and Michel Lussault (Paris: Belin, 2003).

33. And in my opinion, the absence of uniqueness does not lead to the idea of the eternal return of Friedrich Nietzsche or cycles, whatever they are.

34. "Les théories ne procèdent jamais des faits. Les théories ne procèdent que de théories antérieures souvent très anciennes. Les faits ne sont que la voie rarement droite, par laquelle les théories procèdent les unes des autres. Mais à la fois il est difficile de concevoir qu'une théorie puisse entièrement éliminer une autre à son profit." George Canguilhem, *La connaissance de la vie* (Paris: Vrin, 1992), 62.

35. *Ambition* 13 (2007).

36. As explored in the exhibition Non-Standard Architecture, curated by Frédéric Migayrou and Zeynep Mennan at the Centre

Roussel are part of the bachelor machines corpus delineated by Michel Carrouges or Harald Szeeman.

49. *Amazing Stories*, the first science fiction magazine, was first published in 1926, in New York City.

50. *Erewhon* is a utopic novel of Samuel Butler published in 1872. Erewhon is an anagram of "nowhere." Alternately, Deleuze and Guattari suggest it's not only "nowhere," but also "now here."

51. See John Z. Langrish, "Darwinian Design: The Memetic Evolution of Design Ideas," *Design Issues* 20, no. 4 (Autumn 2004): 4–19.

52. Karel Capek's *R.U.R.* (*Rossum's Universal Robots*) was written in Czech in 1920 and performed on Broadway the following year. Capek's brother Josef invented the term *robot*, from the Czech *robota* (worker).

53. The MacGuffin is a concept created by Alfred Hitchcock. It's an item that is used to trigger intrigue. The prominence of the MacGuffin typically decreases in relation to the rising action of the plot. It is a pretext, an alibi to create the artifact that becomes a movie.

54. Corps sans Organe, or CsO, is a concept developed by Deleuze and Guattari in *Mille Plateaux* and *Anti-Œdipe*. Here, it comes from Antonin Artaud, *Œuvres complètes*, vol. 13 (Paris: Éditions Gallimard, 1956), 104. "Man is sick because he is badly constructed. We must decide to strip him in order to scratch out this animalcule which makes him itch to death, god, and with god his organs. For tie me down if you want to, but there is nothing more useless than an organ. When you have given him a body without organs, then you will have delivered him from all his automatisms and restored him to his true liberty. So, you'll learn to dance in reverse as in the delirium of the bal-musette, and this will be its true place."

55. Fernando Pessoa, "Anarchism," *Magazine Littéraire* 291 (September 1991): 29.

56. Felix Guattari, *Les trois ecologies* (Paris: Éditions Galilée, 1989).

57. Rudolph Steiner, *Anthroposophische Leitsätze* (Dornach, Switzerland: Philosophisch-Antroposophischer Verlag, 1925).

58. Gilles Deleuze and Michel Foucault, *Présentation de Sacher Masoch: Le Cruel et le Froid* (Paris: Éditions de Minuit, 1967).

59. Michel Foucault, *Utopie et Heterotopie*, audiotape of lecture by Foucault for National French Radio, Paris, 1966.

60. Comment by B: Of course the result of the

13. Only metaphors of the State as a master of wars, Pygmalion of arts or benefactor of widows and orphans... We can only look at them with an epistemologic tenderness.

14. And by the way, couldn't we consider the human brain the most sophisticated virus that ever existed on Earth?

15. See Eyal Weizman, "Lethal Theory," *Log* 7 (2006).

16. See, for instance, David Usborne, "Toxic waste creates hermaphrodite arctic polar bears," the *Independent*, 10 January 2006.

17. See Steven Johnson, *Emergence* (New York: Scribner, 2002).

18. Bernard Rudofsky, *Architecture Without Architects: An Introduction to Nonpedigreed Architecture* (New York: Museum of Modern Art / Garden City, NY: Doubleday, 1964).

19. Frédéric Migayrou in *Architecture Non Standard*, the catalog of the exhibition, Centre Pompidou, 2004.

20. Both films were directed by the notorious Stanley Kubrick, whose first feature was *Fear and Desire* (1953), depicting a team of soldiers trapped behind enemy lines in a fictional war.

21. Guy Debord, *Commentaires sur la société du spectacle* (Paris: Buchet/Chastel, 1967); Constant and Guy Debord, "La déclaration d'Amsterdam," *Internationale situationniste* 2 (December 1958): 31–32.

22. Frederic J. Schwartz, *The Werkbund: Design Theory and Mass Culture Before the First World War* (New Haven, CT: Yale University Press, 1996).

23. J. G. Ballard, "Back to the Heady Future," *The Daily Telegraph*, April 17, 1993.

24. [Pierre-Simon] Laplace's Demon, named after the eighteenth-century French scientist who devised it, was able to know, at any given moment, all the parameters of the particles of the universe. This first demon was followed by others, such as [James Clerk] Maxwell's Demon, described at the end of the nineteenth century.

25. About heterarchies, multiple disorders, and scenario: there is a part of the brain called the hippocampus. Depressed people have fewer neurons there and fewer interconnectivities between them in this part of the brain. It's only recently that we have been able to scan the hippocampus's activity to understand the effect of different therapies. The fact is that both drugs and traditional psychotherapy seemingly allow the hippocampus to reconstitute its neurons. The ramifications are quite big. To say it fast, it would mean that for Plato, St. Augustine, and all Cartesian and objective schools of philosophy, their distinction between the >

Pompidou, Paris, October 12, 2003–January 3, 2004.

37. See mention of Joseph-Louis Lagrange (1736–1813) in Isabelle Stengers, "La Guerre des sciences," in *Cosmopolitiques 1* (Paris: La Découverte, 1997).

38. Ibid.

39. Ace, 1955. By the way, the end of Philip K. Dick's books are always pretty deceptive, like he didn't know how to finish them; but we can also think that it was only another process of endlessnesslessing his fiction.

40. The Universal Turing Machine is an abstract device conceived by Alan Turing in 1936. Turing used the figurative machine to demonstrate a solution to the following question posed by mathematician David Hilbert: Would it be possible to define a method that could prove the validity of all mathematical affirmation? Maybe one day we will use such non-deterministic machines when we have access to theoretical "DNA computers."

41. Antonin Artaud, "Pour en finir avec le jugement de dieu," audiotape of November 1947 broadcast for French National Radio.

42. Marcel Duchamp, audiotape of interview by Alain Jouffroy, *Rencontre*, 1961. Available online at http://www.ubu.com/sound/duchamp.html.

43. The Centre d'Etudes de Mathématique et Automatique Musicales (CEMAMu), founded by architect/composer Iannis Xenakis, completed its Unité Polyagogique Informatique du CEMAMu (UPIC) program in 1977. Able to express the "sound" of shapes, UPIC created synergy between math and music. Xenakis conceived of UPIC while working on the Philips Pavilion in 1958; the pavilion's hyperbolic paraboloids having been inspired by his 1955 musical composition, *Metastasis*.

44. Gilles Deleuze and Félix Guattari, *L'anti-Œdipe*, vol. 1, *Capitalism et Schizophrenie* (Paris: Éditions de Minuit, 1972).

45. Robert Linhart, *L'etabli* (Paris: Éditions de Minuit, 1978). Linhart came from a group of Trotskyist and Maoist students who, after the May '68 strike in France, decided to stop studies and infiltrate Citroen as unqualified workers to test their ability to resist inside the capitalist system itself. The experiment was published as a book in 1978.

46. J. G. Ballard, "Which Way to Inner Space," in *New Worlds* (1962).

47. Auguste Villiers de L'Isle-Adam, *L'Ève future* (1886).

48. The famous books *Impressions d'Afrique* (1910) and *Locus Solus* (1914) by Raymond >

apparatus equation is not equal to the sum of all those ingredients, it has more to do with an alchemical process where the ingredients, when put in contact, transform themselves.

credits

< i've heard about >
authors: R&Sie(n) / François Roche, Stéphanie Lavaux, Jean Navarro / & Benoît Durandin
with: Behrokh Khoshnevis (contour-crafting process, USC, Los Angeles, United States); François Roustang (hypnosis specialist, Paris, France); Chris Delaporte (film director, 3-D effects, Paris, France); Christophe Berdaguer & Marie Péjus (artists, Marseille, France); Mathieu Lehanneur (designer, Paris, France); Laurent Genefort (science fiction writer, Paris, France); CNRS, Laboratoire de Spectrométrie (nanoparticles, Grenoble, France); M/M (graphic designer, Paris, France); Gilles Schaeffer (mathematician, Paris, France); Michel Boulcourt (landscape architect, Paris, France); Alexandra Midal (art critic, Paris, France); Matthieu Kavyrchine (video artist, Paris, France); Sébastien Szczyrk (sound designer, Paris, France); Alexandre Merlet (video producer, Paris, France); Stephan Heinrich (architect, Berlin, Germany); Providence (singer, Paris, France)
prototype / installation / publishing: Ufacto, David Toppani (prototype 1:1 scale model, Paris, France); Onestar Press (neighborhood protocol publishing, Paris, France); Christian Hubert de Lisle (prototype, Montpellier, France); Thibaut Boyer (3-D computing shape, Vouzon, France)
sponsors, partnership: Musée d'Art Moderne de la Ville de Paris (France); MUDAM (Luxembourg); deSingel (Antwerp, Belgium); USC (Los Angeles, United States); 21st Century Museum of Contemporary Art (Kanazawa, Japan); CNC/DICRéAM (Paris, France); Lafarge (Paris, France); Materialise (Leuven, Belgium); Next Limit Technologies (Madrid, Spain); Dapa (Choisy-le-Roi, France); University of Architecture (Innsbruck, Austria); New-Territories (Paris, France); CNRS (Grenoble, France)
production: Paris-musées (France)
curators: Laurence Bossé (France), Angeline Scherf (France), Hans Ulrich Obrist (Switzerland)
exhibitions: 21st Century Museum of Contemporary Art (Kanazawa, Japan); Musée d'Art Moderne de la Ville de Paris (France); deSingel (Antwerp, Belgium)

< hypnochamber >
authors: R&Sie(n) / François Roche, Stéphanie Lavaux, Jean Navarro / & Benoît Durandin
with: Mathieu Lehanneur (seat-shell design, Paris, France); François Roustang (hypnosis specialist, Paris, France)
exhibition: Musée d'Art Moderne de la Ville de Paris (France)

< green gorgon >
authors: R&Sie... / François Roche, Stéphanie Lavaux, Jean Navarro / & Philippe Parreno
with: Guscetti & Tournier (engineering, Geneva, Switzerland); VP & Green (envelope engineering, Paris, France); Klaus Daniels, HL-Technik (climatization engineering, Munich, Germany); Michel Boulcourt (landscape architect, Paris, France); Mark Dion (artist, New York, United States); Mathieu Lehanneur (GPS designer, Paris, France); Sergio Ochatt (botanist, Paris, France); Ami Barak (curator, Paris, France)
collaborators: Miguel-Angel Munoz, Quck Zhong-Yi, Kika Estarellas, Maud Godard, Julien Jacquot

< olzweg >
authors: R&Sie(n) / François Roche, Stéphanie Lavaux, Jean Navarro
with: Pierre Huyghe (artist, Paris, France / New York, United States); Mathieu Lehanneur (furniture designer, Paris, France); Stephan Heinrich (robot designer, Berlin, Germany); Nicolas Green (facade engineer, Paris, France / London, United Kingdom); Ami Barak (curator, Paris, France); Sibat (basic engineer, Paris, France); Julien Blervaque (script programmer, Montreal, Canada)
collaborators: Alexander Römer, Agnes Vidal, Daniel Fernández Florez

< loophole (second step 2008) >
authors: R&Sie(n)... / François Roche, Stéphanie Lavaux / & Marc Fornes (architect, clip weaving script, New York, United States)
with: VP & Green (engineering, Paris, France)
collaborators: Sylwia Bogdan, Toshikatsu Kiuchi

< shearing >
authors: R&Sie... / François Roche, Stéphanie Lavaux, Alexandre Boulin, Olivier Legrand
with: Abaca (engineering, Montpellier, France); Christian Hubert de Lisle (contractor, Montpellier, France)

< spidernethewood >
authors: R&Sie(n) / François Roche, Stéphanie Lavaux, Jean Navarro
with: Nicolas Green (engineer, Paris, France / London, United Kingdom)

< he shot me down >
authors: R&Sie(n) / François Roche, Stéphanie Lavaux, Jean Navarro
machines: Stephan Heinrich
with: Leopold Lambert

< mi(pi) bar >
authors: R&Sie(n) / François Roche, Stéphanie Lavaux, Jean Navarro
with: Julien Blervaque (Voronoi geometric script, Montreal, Canada)

< mosquito bottleneck >
authors: R&Sie... / François Roche, Stéphanie Lavaux, Jean Navarro, Pascal Bertholio
with: Mathieu Lehanneur (designer, Paris, France)

< aqua alta >
authors: Roche, DSV & Sie / François Roche, Stéphanie Lavaux, François Perrin, Ammar Eloueini
collaborators: Martial Vieille, David Toppani

< b_mu/dusty relief >
authors: R&Sie... / François Roche, Stéphanie Lavaux, Jean Navarro, Pascal Bertholio
with: A49 (local architects, Bangkok, Thailand); Michel Boulcourt (landscape architect, Paris, France); Mathieu Lehanneur (furniture designer, Paris, France); Nicolas Green (structural steel engineer, Paris, France / London, United Kingdom); ACT Espace / Benoît Lalloz (light engineer, Paris, France)

< waterflux >
authors: R&Sie(n) / François Roche, Stéphanie Lavaux, Jean Navarro
with: Guscetti & Tournier (engineering, Geneva, Switzerland)
Museum apparatus: Mathieu Lehanneur

< hybrid muscle >
authors: R&Sie... / François Roche, Stéphanie Lavaux, Jean Navarro / & Philippe Parreno (The Game; Boys from Mars)
with: Christian Hubert de Lisle (contractor, Montpellier, France), ADS (machine contractor, Paris, France)

< terra incognita >
authors: R&Sie(n) / François Roche, Stéphanie Lavaux, Jean Navarro / & Pierre Huyghe
with: Julien Blervaque (script, Montreal, Canada)
collaborators: Camille Lacadee, Clarisse Labro
exhibitions: Park Celebration by Pierre Huyghe, Musée d'Art Moderne de la Ville de Paris (France), March–April 2006; Tate Modern (London, United Kingdom), July–August 2006

< swarm town >
authors: R&Sie(n) / François Roche, Stéphanie Lavaux, Jean Navarro / & Benoît Durandin, Petra Jenning, Florian von Hayek, Alain Marder-Greco, Eglantine Bigot-Doll

< thegardensofearthlydelights >
authors: R&Sie(n)... / François Roche, Stéphanie Lavaux
with: Barbara Ozimec (botanist, Canada)
collaborators: Leopold Lambert, Jakob Ingemansson

< (un)plug >
authors: R&Sie. D/B:L / François Roche, Stéphanie Lavaux, Alexandre Boulin, Olivier Legrand, Gilles Desevedavy
with: Benoît Durandin, Étienne Feher, David Toppani

< i'mlostinparis >
authors: R&Sie(n) / François Roche, Stéphanie Lavaux, Jean Navarro
with: Christian Hubert de Lisle (structural development and construction, Montpellier, France)
glass beakers: Pedro Veloso, after consulting Vanessa Mitrani

R&Sie(n) is an architecture office based in Paris and headed by François Roche (born 1961, graduated from UPA 3, Versailles, 1987), with Stéphanie Lavaux (born 1966, studied at the ENS des Beaux-Arts until 1990). Set up in 1989, the office has changed names several times: BoyeRoche, 1989; Roche*, 1990; Roche&François, 1991; Roche, François, Lewis, Huber, Roubaud, Perrin, 1992; Roche, DSV & Sie, 1993–97; R, DSV & Sie .P, 1998; R&Sie.D/B:L, 1999–2001; R&Sie…, 2001–4; R&Sie(n), 2004–8.

Their research into a critical experience of architecture occurs through a mutation of contextual parameters. Scenarios of hybridization, grafting, cloning, and morphing give rise to perpetually transforming architecture that strives to break down the antinomies between object/subject or object/territory. Experimental and inventive, the architecture of R&Sie(n) seeks to be profoundly critical; an architecture that is articulated as multiple narrative apparatus and relationship strategies, made of substances from each situation.

The organic, oppositional architectural projects of R&Sie(n)'s practice are concerned with the bond between building, context, and human relationships. They use speculations and fictions as a process to disalienate post-capitalism subjectivities. R&Sie(n) consider architectural identity to be an unstable concept, defined through temporary forms in which the vegetal and biological become a dynamic element. R&Sie(n) are currently experimenting with new warping technologies to prompt architectural "scenarios" of cartographic distortion, substitution, and genetic territorial mutations to create protocols in apparatus that merge fiction, subjectivization, and transformation processes with "bachelor and desirable" machines, simultaneously re-scenarizing the aesthetic relationship with nature(s)—artificial, paranoid, and/or real.

R&Sie(n)'s projects have been exhibited in institutions such as the Tate Modern (London, 2006), Columbia University (New York, 1999–2000), UCLA (Los Angeles, 1999–2000), the ICA (London, 2001), the Mori Art Museum (Tokyo, 2004), Centre Pompidou (Paris, 2004), the Musée d'Art Moderne de la Ville de Paris (2005, 2006), MIT's Media Lab (Cambridge, 2006), Pavillon de l'Arsenal (Paris, 2006), and the Orléans/ArchiLab International Architectural Conference (1999, 2001, 2003). R&Sie(n) were selected to take part in the French exhibitions at the Venice Biennale in 1990, 1996, 2000, and 2002 (refused invitation), and to the international exhibition in 2000, 2004, and 2008.

R&Sie(n) has held several professorships through François Roche—at the Bartlett School of Architecture (London), TU (Vienna), ESARQ (Barcelona), ESA (Paris), UPenn (Philadelphia)—and is now teaching an advanced studio at Columbia (New York). R&Sie(n) has lectured at MIT, Harvard, the AA, UCLA, and Sci-Arch, among other universities.

Benoît Durandin
has been part of R&Sie(n) since 1999. He recently worked on the project I've Heard About, an experiment for a self-growing, urban entity. This project, started in 2003, has exhibited in Kanazawa, Paris, Antwerp, and Tokyo. He lives and works in Athens, Greece, where he founded MO Y, an architecture and research office, with Katerina Chryssantopoulou in 2008. He is the editor of *Multitudes*, a magazine about art, philosophy, and politics.

Alexandra Midal
is a professor in the theory and history of design at ECAL, in Lausanne, Switzerland, and at the École Supérieur des Beaux-Arts in Toulouse, France. She is the former director of the Fonds Régional d'Art Contemporain in Haute-Normandie and an independent curator. She has published many books and catalogues, including *Appartement témoin, les années 50* and *Appartement témoin, les années 60*, *Antidesign: petite histoire de la capsule d'habitation en images*, and, in 2008, *Tomorrow Now: When Design Meets Science Fiction*.

Joker
is an impertinent and provoking character.